BISCUIT MANUFACTURE

BISCUIT MANUFACTURE
fundamentals of in-line production

by

PETER R. WHITELEY

DIP. F.E., F.Inst. B.B.

*Head of the Department of Baking,
Rush Green Technical College,
Romford, Essex*

ELSEVIER PUBLISHING COMPANY LTD

LONDON–AMSTERDAM

1971

ELSEVIER PUBLISHING COMPANY LTD
BARKING, ESSEX, ENGLAND

ELSEVIER PUBLISHING COMPANY
335 JAN VAN GALENSTRAAT, P.O. BOX 211, AMSTERDAM
THE NETHERLANDS

444-20072-X

LIBRARY OF CONGRESS CATALOG CARD NUMBER 70-122961
WITH 121 ILLUSTRATIONS AND 5 TABLES
© 1971 ELSEVIER PUBLISHING COMPANY LIMITED

All rights reserved. No part of this publication may be reproduced, stored in a retrieval system, or transmitted in any form or by any means, electronic, mechanical, photocopying, recording, or otherwise, without the prior written permission of the publishers, Elsevier Publishing Company Ltd, Ripple Road, Barking, Essex, England.

Set in cold type by Peridon Ltd., London, NW9
and printed in Great Britain by Redwood Press, Trowbridge, Wiltshire

Author's preface

THE intention of this book is to provide a guide for potential management and supervisors and for those who wish to understand the fundamental principles of biscuit manufacture. It does not set out to be a learned treatise. The purpose of the book is to simplify and explain processes and materials so that the 'mystique' is replaced by logic. Once the mystique is removed the biscuit maker is one step closer to anticipating and solving problems.

In attempting to cover this subject within one concise volume, it is difficult to avoid over-simplification or generalisation, and apologies must be offered in advance where these occur. To wallow in the fine details of specialisation is to defeat the object of the book, and less would be achieved if the issues were confused. The reader's attention is drawn to the interpretation of formulae (recipes). Raw materials, equipment, methods, processes, and conditions vary considerably; the formulae are intended as blue prints from which, with a knowledge of the materials and aims of the processes, and by trial and error, a biscuit can be produced bearing some semblance to the original. All formulae should be interpreted in conjunction with the 'Guide to using formulae' at the beginning of Chapter 12.

As the biscuit industry advances towards complete automation, plant and equipment become more advanced and sophisticated. Training of a very high standard is necessary to use sophisticated machinery efficiently, and where skills of this exacting degree are not available, then the simpler, less advanced equipment, should be used. It is with this in mind that processes have been dealt with in such a way that they can be carried out by hand or through varying degrees of mechanisation.

Similarly, the legislation regarding foods becomes so involved and changes so frequently that it is impossible to review for inclusion in a book of this nature. It is essential to be aware that there is legislation pertaining to practically every aspect of biscuit making and marketing, and as the legislation differs from country to country, further confusion arises. All governments issue pamphlets about their laws, and the appropriate Ministry or information office (in the United Kingdom, Her Majesty's Stationery Office) or legation should be approached for the current regulations.

Upminster, 1970   PETER R. WHITELEY

Acknowledgements

AN expression of thanks and gratitude is due to all those who helped with advice, information and photographs to make this book possible, including:

Baker Perkins Ltd, Peterborough
British Cellophane Ltd, Twickenham
Co-operative Wholesale Society Ltd, Biscuit Factory, Harlow
Walter Denis Contacts Co., Ltd, Blackpool
Flour Milling and Baking Research Association, Chorleywood
J. Alan Goddard Ltd (Hecrona), Croydon
J. W. Greer Co., Ltd, Bromley
Kek Ltd., Macclesfield
A. M. Lock & Co. Ltd, Oldham
Machinery Continental Packaging Ltd (Aucouturier), Rochester
Morton Machine Co. Ltd, Wishaw
E. T. Oakes, Ltd, Macclesfield
Radyne Ltd, Wokingham
Rank Precision Industries Ltd, Brentford
Rose Forgrove Ltd, Leeds
SIG Wrapping Machines Ltd, Croydon
Simon-Vicars Ltd, Newton-le-Willows
Spooner Food Machinery Engineering Co., Ltd, Ilkley

My greatest appreciation is due to my wife, whose untiring efforts of encouragement and help spurred me on at all times.

Upminster, 1970   PETER R. WHITELEY

Contents

AUTHOR'S PREFACE v
ACKNOWLEDGEMENTS vii
LIST OF FIGURES IN THE TEXT xi
LIST OF TABLES xii
LIST OF PLATES xiii

PART I RAW MATERIALS

1 Flour and cereal products 1
2 Fats and oils 22
3 Sweetening agents 34
4 Aerating agents 42
5 Dairy products 50
6 Fruits and nuts 59
7 Setting materials 69
8 Chocolate and cocoa products 73
9 Flavouring materials 81
10 Colouring materials 94

PART II CLASSIFICATION AND METHODS

11 Classification of biscuit types and methods of production 103

PART III FORMULAE–QUALITY CONTROL AND DEVELOPMENT

12 Basic ingredient proportions of biscuit doughs 127
13 Basic ingredient proportions of wafers, marshmallows, creams, and fillings 148

14	Quality control	161
15	Re-use and disposal of unsatisfactory products	189
16	Development	193

PART IV   PLANT AND EQUIPMENT

17	Raw materials storage and handling	201
18	Mixing room equipment	209
19	Machine room equipment	219
20	Ovens and baking	230
21	Wafers and second process equipment	241
22	Ancillary equipment and automation	250
23	Packaging of biscuits	258

PART V   GENERAL CONSIDERATIONS

24	Factory layout and hygiene	269

BIBLIOGRAPHY   289

INDEX   290

List of figures in the text

1. Longitudinal section of a wheat grain — 4
2. The pH scale — 47
3. Sketch of a yeast cell — 106
4. Yeast reproducing by budding — 107
5. Colony of yeast cells — 108
6. Diagram showing enzymic activity during fermentation — 111
7. Structure of a farinogram — 163
8a. Farinogram: Manitoba flour — 163
8b. Farinogram: strong English flour — 164
8c. Farinogram: soft English flour — 164
9a. Extensogram: Manitoba flour — 166
9b. Extensogram: strong English flour — 167
9c. Extensogram: soft English flour — 167
10. Typical alveograms: (a) Manitoba flour; (b) strong English flour; (c) soft English flour — 168
11. Typical graphs of extensometer: (a) Manitoba flour; (b) strong English flour; (c) soft English flour — 170
12. Line drawing of Simon-Vicars vertically integrated dough feeding, gauging and laminating unit — 221
13. Examples of wire band patterns (Baker Perkins) — 231
14. Air blast gas burners for direct fired oven (Baker Perkins) — 235
15. Diagrams illustrating the principle of forced air convection baking used in a Spooner oven section — 236
16. Line drawing of a Walden refrigerated air blast cooling tunnel — 248
17. Diagrams illustrating how metals affect the electromagnetic field of an electronic metal detector — 254

List of tables

1. Structural composition of wheat   3
2. Typical analyses of flour samples   9
3. Typical analyses of milk and milk products   53
4. Table converting ingredient percentages to pounds when based on flour weight of 280 lb (i.e. one sack) and vice versa   128/129
5. Examples of texture meter readings obtained on retail samples of various types of biscuits   181

List of Plates

(between pages 140 and 141)
1a. Shortcake fingers
1b. Round shortcake biscuits
2a. Shortcake biscuits: 'Royal Duchess'
2b. Lincoln (Lincoln Creams)
3a. Nice
3b. Finger creams
4a. Custard creams
4b. Bourbon creams
5a. Digestive (Sweetmeal)
5b. Currant biscuits
6a. Gingernuts
6b. Half-coated (chocolate) sweet biscuits
7a. Coconut cookies (wire-cut)
7b. Cookies with chocolate, nuts and currants included
8. Cream crackers
9a. Savoury crackers
9b. Creamed puff shells
10a. Rich Tea biscuits
10b. Morning Tea biscuits
11a. Marie biscuits
11b. Finger-shaped, semi-sweet, hard dough biscuits
12a. Garibaldi biscuits
12b. Morning tea biscuit showing hair-line fracture known as 'checking'
13a. Coconut mallows
13b. Chocolate teacakes
14a. Jam rings
14b. Fig bars
15a. Cream filled wafers
15b. Bag-type pack of coconut cookies
16a. Tray-type pack for mallows using a preformed liner
16b. Tray-type pack for assorted biscuits using a preformed liner
    (between pages 204 and 205)
17a. Brabender Farinograph

LIST OF PLATES

17b. Brabender Extensograph
18a. Chopin Alveograph
18b. Simon Research Testing Unit
19a. Baker Perkins Texture Meter
19b. Baker Perkins Texture Meter showing interior
20a. Douglas bulk fat storage vessels
20b. Douglas fat processing equipment with mixing tanks on the right and the emulsifier-cooler on the left
21a. Douglas silos for storage of plasticised blended fats
21b. Control room for bulk handling of ingredients
22a. Syrup blending tanks
22b. Morton HD 120 three-speed vertical mixer of the planetary type
23. Simon-Vicars two-spindle vertical mixers
24a Morton Duplex 4½ horizontal 'Z' blade mixer
24b. Morton Gridlap GL 70 horizontal mixer with control console
25a. Simon-Vicars high-speed mixer showing interior
25b. Simon-Vicars high-speed mixer ejecting biscuit dough directly into a floor mounted hopper
26a. The Oakes continuous mixer/modifier for bread and biscuit doughs
26b. Mixing rotor shaft of the Oakes continuous mixer/modifier
27a. The Oakes continuous automatic mixer for batters, marshmallow and fluid mixings
27b. Stator and rotor of mixing head of the Oakes continuous automatic mixer
28a. Morton 100 two-speed air pressure whisk
28b. Simon-Vicars floor mounted tub discharger
29a. Simon-Vicars tub elevator and discharge unit to dough feeder

LIST OF PLATES

29b. Simon-Vicars dough feed to sheeter showing scrap return
30a. Morton heavy duty reversing dough brake
30b. Simon-Vicars right-angle laminator
31. Hecrona vertical laminator
32. Simon-Vicars· cutting machine showing dough feed, pre-sheeter, three pairs of precision gauge rolls and cutting machine cross head
33. Baker Perkins rotary cutting unit showing the pressure roller situated beneath the impression and cutting rollers
34a. Simon-Vicars rotary cutting unit in use with the guards removed
34b. Simon-Vicars fig bar extrusion unit situated directly above the oven band
35. Baker Perkins 88 BT rotary moulder with dough feed
36. Baker Perkins 88 BT rotary moulder with hopper opened showing the moulding roller and the grooved forcing roller
37. Baker Perkins Turboradiant oven with direct-fired boost on the first section
38a. Oil-fired heat sources for one section of a Baker Perkins Turboradiant oven
38b. Tower-type tensioning for oven band (Simon-Vicars)
39. Spooner forced air convection oven with wire band
40. Oven control panel for a Simon-Vicars gas-fired oven (in the background can be seen a radial reverse turn carrying the biscuits to an overhead cooling conveyor)
41. Radyne OCB 301 electronic baking unit
42. Simon-Vicars Mark II wafer oven
43a. Simon-Vicars wafer sandwich building machine with creamer in the background
43b. Simon-Vicars wafer saw
44. A battery of Simon-Vicars quality creamers

being fed directly from the cooling conveyor of the oven
45. Oakes oven pacer depositor producing marshmallow teacakes
46a. Walden automatic chocolate kettle
46b. Walden continuous trickle feed and tempering machine
47a. Goddard automatic chocolate tempering machine
47b. The Walden Supreme Mark 7 automatic enrober
48. Baker-Sollich 76CQ enrober, a sophisticated piece of equipment capable of control by computer

(between pages 236 and 237)

49a. Walden refrigerated air blast cooling tunnel
49b. Peck Mix for biscuit crumbing and raw materials sieving
50. Kek Ltd. laboratory-size grinding mill showing the discs and arrangement of the pegs
51a. Simon-Vicars Mark III oil spray unit with cover removed to show interior
51b. Close up of the Baker Perkins oil spray unit showing the oil dispersers above and below the wire band
52. Greer spiral conveying system
53a. A Lock electronic metal detecting unit monitoring fully coated wafer sandwiches
53b. An example of automatic control by Baker Perkins Developments Ltd. (the instrument in the top centre of the photograph, by means of sensitive feelers, keeps the biscuit rows and the stacker guides in constant alignment)
54. Aucouturier wrapping machine, fed direct from the cooling conveyor, for round biscuits

LIST OF PLATES

55. Aucouturier wrapping machine for round packets with end seals
56. Aucouturier wrapping machine hand fed for square biscuits in piles using corrugated grease-proof paper and a heat-seal overwrap (mode of wrapping is also illustrated)
57. Rose Forgrove wrapping machine for pillow-type packages
58. 'Sendform' thermoforming machine for producing moulded plastic packing trays
59. SIG roll wrapping machine with automatic feeder
60. Close up of a SIG automatic loader (type ZH)
61. SIG automatic wrapping machine for fin-sealed packets with automatic feeder
62. SIG automatic wafer wrapping machine
63. Jones-Rose constant motion cartoning machine
64. Rose Forgrove bag forming and filling machine for tumble packs

PART I
Raw materials

CHAPTER 1

Flour and cereal products

FLOUR is the basic raw material of biscuit production and probably the most variable. It is therefore extremely important to know something of its origin and production in order to understand the handling and manufacturing of biscuits.

WHEAT

In the Western world, and those parts under Western influence, wheaten flour is the basic food for existence. In other parts, different cereals are prominent, but they are only of minor importance in this context.

Wheat is a farinaceous grass, known botanically as *triticum vulgare*. It is thought to have originated in the Middle East, but it is unlikely that those specimens would closely resemble the highly developed wheat of today. Development has produced good yielding varieties and strains suitable for prevailing climates ranging from near the Equator almost to the Arctic circle.

Even within this range of conditions best results are obtained on heavy loamy soil, with moderate rainfall in the growing season, and intermittent light rain during a hot summer; wheat will, however, flourish on light soil or in a cool temperate climate.

The climate and soil have a marked influence on the type of wheat grown and subsequently on the flour produced. Other major influences can be summed up as follows:

(1) place of growth
(2) time of planting
(3) colour of grain.

Place of growth

The principal growing countries are Russia, China, and the USA, but as these consume the majority of their production they have little, if any, to export. The major exporting countries of wheat are Canada, Argentine, and Australia; France and Great Britain are considerable producers, but, Britain, in particular, is a considerable consumer.

Time of planting

(a) Winter: The grain is planted in the autumn, germinates and sprouts to a height of 3 or 4 in (75-100 mm) before the winter. Growth is arrested until spring when it is able to start growing long before land is suitable for sowing. The winter conditions must not be too severe or the plants will suffer frost damage. The wheat is frequently insulated from the hardest frosts by a layer of snow.

This is the most common type of wheat grown in Great Britain, as the winters are long and wet and the land becomes too heavy for tractors and equipment to sow sucessfully in the early spring.

(b) Spring: The grain is planted in the spring, where the climate is suitable for speedy growth and ripening, as the plant must reach maturity in a few months only. This climate exists in Canada where, in fact, the winters are so severe that it is doubtful if autumn-sown wheat would survive.

Colour of grain

The colour of grain depends upon pigment in the bran, and is usually divided into 'red', 'yellow', and 'white' wheat. The pigmentation usually depends upon type of wheat, soil and climate, and is generally a good guide to character.

Wheat and flour characteristics

Strains sown in winter usually yield a soft wheat, and consequently, a soft or weak flour; similarly, white wheats are usually soft. These are the types normally grown in Britain, and English flour is usually a weak flour, entirely satisfactory for the production of soft-dough biscuits. Yellow wheats tend to yield medium strength flours.

Spring-sown wheats produce a hard grain from which is produced a strong flour, and red wheats are also usually hard. For example, a red, spring Manitoba wheat would produce an extremely strong flour, suitable only for long fermentation processes, and possibly for puff doughs. The latter would probably need weakening with a proportion of English flour.

Thus it will be seen that Canadian, Russian, and North USA wheats, are normally hard red spring wheats, producing very strong flours, suitable for long fermentation processes only.

The USA also produces soft wheats for soft and hard-dough biscuits. In addition, a flinty hard wheat, known as Durum, is Produced in the USA, which has little use in milling, but is used in the production of semolina, macaroni, and spaghetti.

The Argentine wheats, known as 'Plate' wheats, produce a medium-strength flour which is used largely by the miller for blending purposes.

Imported Australian wheats are soft, and as they have a very thin bran layer, they yield a high proportion of flour. The flour is soft and has good creamy colour and flavour.

English and French wheats are soft, producing a soft flour of good colour and flavour, and are probably the best for soft- and hard-dough biscuits.

From the foregoing factors it will be realised that in many ways the wheat grain varies both physically and chemically. Physically, it varies in shape, size, and colour, and chemically, in the type of flour that can be produced from different grain. Wheat also varies in its composition, but the following table will give a general idea of the make-up of the grain, and the diagram will help in showing its physical structure (Fig. 1).

TABLE 1: *Structural composition of wheat*

	%	
Epidermis	0·5	
Epicarp	1·0	
Endocarp	1·5	
Testa (Episperm)	2·0	Bran = 13·0%
Nucellar layer	1·0	
Aleurone layer	7·0	
Germ	2·0	
Endosperm	85·0	
	100·0	

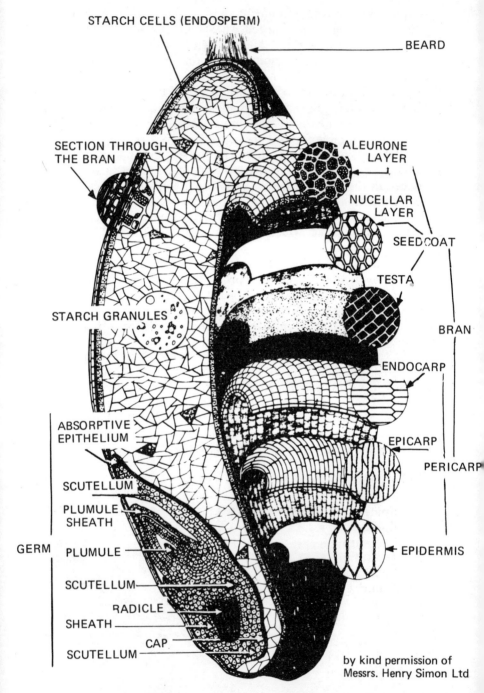

Figure 1. Longitudinal section of a wheat grain

The bran: The skin of the wheat grain is known as bran and consists of six different layers, each of which is composed chiefly of cellulose. Cellulose is not only tough and fibrous, but it is also undigestible by the human alimentary system. It does, however, play an important part as roughage and bulk. It is invaluable as cattle food, cattle being equipped to digest bran. The first three layers form the proper skin of the grain, and the layers lie in different directions for strength, rather like the sheets of wood used in plywood. The fourth layer, known as the testa (and also the episperm), is the true envelope of the grain, and as it contains the pigment, it is responsible for the colour of the grain, and very closely associated with the almost waterproof nucellar layer. The innermost layer is really part of the endosperm, but because of its high protein and fat content, it is undesirable to retain in the flour, and is recovered in milling with the bran.

The germ: The germ is pale yellow in colour, having a high proportion of protein, fat, and enzymes, and is devoid of any starch. It is built up of three component parts, each having specific duties in the growth of a new plant.

(a) *the plumule,* which consists of the undeveloped leaves of the future plant,
(b) *the radicle,* which is the root centre, and
(c) *the scutellum,* which is the so-called conveyor system, whereby, with the assistance of enzymic action, the endosperm is broken down into simpler and suitable foodstuffs for the initial growth and development of the young plant.

The enzyme cytase breaks down the cellulose envelopes and the starch cell walls, permitting the starch in turn to be broken down into dextrins and maltose by the group-enzyme diastase. The inclusion of activated germ or wheat in fermented doughs can have disastrous results, as the aforementioned sequence can occur producing doughs which are quite unmanageable.

The endosperm: The endosperm is the floury portion of the grain from which the flour is eventually produced. This is composed chiefly of starch. The starch occurs in cell form in two sizes, the cell wall of which is cellulose; packed around the starch cells are the granules of protein—both soluble and insoluble. The endosperm also contains sugars and small quantities of mineral salts.

The soluble proteins are albumen and globulin, and constitute the minor fraction of the total protein content. The insoluble are known collectively as gluten-forming proteins. These are composed chiefly of two proteins, glutenin and gliadin. Glutenin is insoluble except in dilute acids, such as are produced during fermentation, and in dilute alkalis, such as sodium bicarbonate, both of which have a softening effect on gluten. Glutenin is the very tough and fibrous constituent of gluten, and gives gluten its stability. Gliadin on the other hand, which is only soluble in 75% alcohol, has a mellow and elastic nature. The character of the gluten depends upon the proportions of these two components, which in turn have a direct influence on the strength and characteristics of flour.

MILLING OF WHEAT INTO FLOUR

Flour milling is an extremely old art with a very ancient history of evolution and development—from primitive pounding with stones held in the hand to the two systems employed commercially today. The latter are known as stone milling and roller milling.

Stone milling

This method has its roots in Egyptian and Greek history, but is now powered by steam, water, or electricity. Basically, it depends upon grain entering between two stone discs through the central aperture of the upper stone which revolves, generally in contact with the lower stationary stone. The grain is crushed by the stones (which are usually grooved) and passes by centrifugal motion to the outer edge of the stones where it falls out as a meal known as wholemeal. Commercially, this system is only employed for speciality wholemeal flours for which a superior flavour is claimed to be obtained.

Roller milling

This is an exceptionally efficient and highly automatic system producing practically all the flour consumed in this country. It consists of two stages, the first being the grading, cleaning, and conditioning of the wheat ready for milling; the second being the gradual breakdown of the wheat into flours of various grades and offals.

Instead of stones, steel rollers are used to break open the grain. These rollers are of two designs. The first are known as break-rollers, and have grooves or flutings running diagonally across the surface. This is to ensure that the wheat is broken only gradually, and no grinding occurs at this stage. The second type are smooth rollers, known as reduction-rollers, designed to grind the separated endosperm resulting from sieving and aspiration after the grain has passed through the break-roll system. More sieving follows the reduction rolls to produce flour. The unsieved stock is returned to the endosperm stream, and so passes to the next set of reduction rolls for further grinding and sieving.

Flour produced at the first reduction-roll is of the highest quality, and would probably represent 30% of the total wheat milled: this is termed a 30% extraction rate. A 70% extraction rate flour, which is generally considered to be of really good quality, would be a blend of flour from the first two reduction-rolls and probably the third. Extraction rate is usually a guide to quality and colour. The higher the extraction rate becomes the lower becomes the quality and colour. Naturally, the more grain incorporated in the flour, the higher the proportion of bran.

Under the roller milling system, wheatmeal and wholemeal flours are usually produced by blending bran back to a white flour. In this way bran particle size and flour quality can be more readily controlled. The extraction rate of wheatmeal can vary from 85-98%; wholemeal should be 100% extraction, but frequently the germ is left out as it has poor keeping qualities unless treated as in germ-meals. Germ-meals are usually manufactured under proprietary names, but are composed roughly of 75-80% high grade flour (70% extraction) and 20-25% treated germ and bran. The germ is treated either with salt or by heat (or both) to preserve it during storage.

The offals, bran, and grain other than wheat, are used in cattle and poultry foods.

All flours contain additives by Act of Parliament (except wholemeal) and sometimes further treatment by the miller, but these will be discussed later.

Impact milling: or protein displacement process

This is an entirely new concept of milling, and in spite of its high running costs it must have considerable effects upon the present processes and products. The process depends upon the fact that wheat starch cells vary in size, and as the protein clusters round the starch cells it follows that there is a greater concentration of protein around the small starch cells than there is around the large ones. Consequently, if the small cell fraction can be separated from the large cell fraction, then flours will result with a high protein proportion and also with a very low protein proportion. In fact, it is possible to start with a wheat having a 10% protein content and produce flour fractions containing as high as 20% and as low as 5% protein contents.

The system requires a peg mill type machine to give rapid and fine reduction of flour without excessive starch cell damage occurring. The particle sizes are then classified by a combination of aspiration (forced draught of air) and centrifugal force. Aspiration alone is not sufficient. Centrifugal force is necessary to help throw the larger particles to the outer parts of the machine. Three or four fractions can be produced in this way, each of which will have varying proportions of protein and starch, and consequently different characteristics. It should, therefore, be possible for the miller to produce a flour to suit all requirements irrespective of the wheat source.

FLOUR COMPOSITION AND NATURE OF FLOUR

Flour in good condition should be a creamy white powder with a faint, pleasant smell, a slightly sweet taste, and, when squeezed firmly in the hands, should tend to cling together, and yet easily fall apart. When mixed with approximately half its own weight of water it forms a plastic mass known as dough. As has been shown already, the composition of flour can vary considerably, depending upon the different types and sources of wheat used in the grist. The grist is the term for the blend of wheats which the miller selects before milling, in order to produce flour suitable for a specific purpose; Table 2 will illustrate this point:

FLOUR AND CEREAL PRODUCTS

TABLE 2: *Typical analyses of flour samples*

	Flour sample suitable for	
Flour constituents	Soft doughs and semi-sweet hard doughs	Fermented doughs
Starch	74·5	71·5
Moisture	14·0	13·5
Proteins—gluten forming	7·0	10·0
Proteins—soluble	1·0	1·0
Sugar	2·0	2·5
Fat	1·0	1·0
Ash (mineral salts)	0·5	0·5
	100·0%	100·0%

The starch consists of microscopic granules or cells. It is insoluble in water but will absorb moisture through its cell wall, this is the reason why flour increases in moisture content during storage. When heated in water the starch cell expands and the cell wall bursts allowing the starch to come in contact with the water. In this way it dissolves, and when cool it will form a gel or paste, depending upon the concentration.

The soluble proteins in flour are similar to egg albumen, i.e. they dissolve in water and coagulate on heating. An excess of soluble proteins is detrimental to fermented doughs.

The insoluble proteins, or gluten-forming proteins, are most important. Gluten is the structure-forming ingredient in fermented and puff doughs. It not only varies in quantity, but also in quality. For instance, it can be soft and elastic, or tough and rubbery. Gluten only forms when flour is mixed to a dough with water. It can be washed out from a dough by immersion and working in water, to leave a greyish-yellow, sticky, elastic substance, with practically no taste or smell.

The water content of flour can vary from 11% to 15%, depending upon storage conditions and the hygroscopic nature of the starch. If the moisture content goes above 14% there is considerable danger of bacterial and, more particularly, mould growth. From the point of view of cost it is important that the price of flour is not paid for water in excess.

The sugar in flour is chiefly sucrose (cane sugar) and partly maltose (malt sugar), both of which although in only small quantities are important as yeast food for fermented doughs. Otherwise they are unimportant in biscuit making. If the sugar content is higher than 2·5% then it is very likely that the grain has been germinating before milling, and consequently, the enzymes which attack the proteins will be active, the proteins will be unstable, and the flour will be unsuitable for fermented and puff doughs.

The fat in the wheat grain is concentrated in the germ and in the bran, and so a high fat content denotes a low grade flour as far as white flour is concerned.

If flour is completely burned away, a white ash remains. This is the mineral content, and consists of phosphates of calcium, magnesium, and potassium, with traces of iron and aluminium sulphates. Flour additives will increase this figure if of a mineral nature.

In all flours there are traces of fibre or cellulose. As the grade of flour becomes lower, the fibre content becomes higher.

Flour quality

It is obvious that hard and fast rules cannot be laid down regarding flour quality, as apart from different types of flour being required for different types of biscuit, flour quality may need to vary to suit different handling techniques, such as mixing, machining or baking, for quite similar types of biscuits. However, the following simple tests, requiring only a minimum of equipment, will give some indication as to the quality of the flour and also to its suitability. More extensive and positive tests will be dealt with while discussing quality control.

Pekar test for colour comparison: Samples of flour, using one of known quality as control, are placed carefully side by side on a rectangular piece of wood or Perspex measuring approximately 6 in × 2 in (150 mm × 50 mm). The surfaces of the samples are smoothed off to form a continuous face, and where the samples join they should have sharp, clean divisions, without blur or intermixing. The samples are then gently and steadily immersed at an inclined angle into water, and withdrawn. A resting time of

30 min should then pass before examination, when a very true comparison of colour should result. To show up branny particles more distinctly, a 1% solution of pyrocatechol should be sprinkled over the surfaces of the samples at this stage.

Water absorbing properties: Flour samples should be compared with a flour of known quality. Place 28 g flour in a beaker and add distilled water from a burette (10 ml should be safe to start with). Stir into the flour with a glass rod. Add water by 0·5 ml until the correct consistency is achieved. This test requires a certain amount of experience to gauge similar consistencies. The amount of water necessary to form the dough is a guide to the quantity and quality of the gluten present. If a high water absorption results, then it will follow that the flour is a strong one, with a high gluten content of a tough nature: a low water absorption will indicate soft or weak flour. Occasionally, it is possible that a flour with a high gluten content may have very soft or weak gluten, and this test may be misleading, but if followed by a gluten examination, it will be resolved.

Gluten quality and quantity: Take the dough produced in the water absorbing properties test, work well to ensure gluten development, and allow to stand in water for 30 min; work the dough in the water to release the starch and soluble proteins —the liquid may be retained for starch, soluble proteins, and sugar examination (not estimation). The dough must be well worked until there is no starch discoloration of fresh water. Gluten quality can then be noted, i.e. colour, toughness or elasticity, and smoothness. This is probably the most important guide to flour quality by these simple tests. If the gluten is grey, without any creaminess in colour, it is quite possible that the flour has been bleached. If the gluten is rough and shows no desire to form a smooth ball, it is possible that excessive bleaching has occurred, in which case only trouble can be expected. The toughness or elasticity will indicate flour strength, and this, coupled with gluten quantity, is a sure guide to flour suitability.

Where there is no excess water adhering to the gluten it should be weighed and expressed as a percentage of the flour weight. This is the wet gluten percentage, not the dry gluten, which is the more important factor. A rough guide to this is obtained by dividing the wet gluten percentage by three, as hydrated gluten is normally two-thirds water. The correct method is to place the gluten in a drying oven until a constant weight is achieved and then express this as a percentage of flour weight.

A dry gluten percentage for soft doughs of 7-9% is usually desirable, and for fermented doughs a figure of 10-11%, depending upon the length of fermentation period. The longer the fermentation, the higher and stronger should be the gluten, and vice-versa. A Manitoba flour may have a gluten content of 13-15%, which will generally be too high and too strong for any biscuit manufacture, except possibly puff doughs, but it is always available to blend with weaker flours. Although breadmaking flours are designed for fermentation processes they are normally too strong for biscuit making, for which the gluten must be more extensible than for breadmaking, since biscuit dough has to withstand an intense machining in a very short space of time.

Moisture content: Place 5 g of flour in a previously weighed dish and dry in a drying oven until a constant weight is obtained. Calculate loss in weight which equals loss of moisture and express as a percentage of flour weight. Moisture content is important from point of view of storage and cost (cf. notes on composition of flour, pp. 8–10).

Ash content: Place 2 g flour into a previously weighed crucible, heat gently and then strongly until only a white ash is left. Cool in desiccator and weigh. Reheat and reweigh until the weight is constant; calculate the weight of ash and express as a percentage. A high figure will denote either a low grade flour, or the addition of mineral improvers, or a combination of both. If an unsatisfactory result is obtained, further tests will be necessary and these will be dealt with later.

Fermentation tolerance: This is a test that will indicate flour strength and can be carried out in either of the following manners.

(a) Cylinder test: 20 g samples of a known and unknown flour or flours are made into batters or ferments with 30 ml water in which are dispersed 2 g yeast. Each ferment is poured into identical 300 ml graduated cylinders which are maintained at a

constant temperature of 27°C (80°F). The initial level readings are taken at intervals of 5 or 10 min. Deductions can be made from the speed of rise as to the available sugar present in the flour: from the length of time to collapse, and from the height attained, as to the strength of the flour.

(b) Dough test: To a 5·0 g sample of flour add 2·75 ml of a 10% yeast suspension in water and mix to a dough free from loose flour. Knead the dough in the palm of one hand with the thumb of the other hand to a seam-free ball over a period of 55 sec, and drop into a beaker of water in a water bath at a constant temperature of 32°C (90°F). Note the time. After a time the dough ball will rise to the surface and will eventually collapse. At this point note the time again, and the overall time is expressed in minutes as the fermentation tolerance. A very low figure, such as 20, will denote an exceptionally weak flour. Typical figures for English flour will be in the 60-100 range. Figures of over 300 should denote a flour suitable for fermentation. Manitoba wheat flours will give figures greater than 500. The biggest drawback to this test is that the amount of handling given during mixing can vary, but with experienced operators, quite a remarkable degree of reproducibility can be achieved.

FLOUR TREATMENT

To produce a flour according to a particular specification and price, the miller is tempted to resort to the addition of chemicals and treatments calculated to improve the colour, strength, softness, or general handling qualities of a flour which does not come up to the standards required. The biscuit manufacturer, on the other hand, prefers to buy untreated flours, meeting any deficiencies with his own treatment, or by returning the flour to the miller: this latter action usually has a salutary effect on future deliveries.

If flour is stored for a period up to, but not exceeding, 6 months, there is a gradual process of bleaching. This is caused through oxidation of the carotene. Carotene is a yellowish coloured substance present in wheat oil, and is responsible for the creamy colour of flour, but when oxidised by either a reagent or time and oxygen of the air, then it loses its colour completely. All bleaches rely on this action, but few have any effect on the branny particles in the flour. The inclusion of a bleaching agent is to achieve the natural improvement in a short time, or to improve a poor grade flour. Storing of flour also improves the stability of the gluten, possibly through the astringent effect by the increased acidity of flour as it ages. Chemicals are also used by the miller to improve the stability and strength of flour, and these generally rely upon their astringent or toughening effect upon the gluten in the dough.

Owing to the vast quantities of flour being produced and consumed it is not a practicable or economic matter for either the miller or the baker to store flour for more than a matter of 2-3 weeks. Another method of achieving this result without the use of chemicals is to heat flour to 64°C (180°F) for 6 to 10 hr, and treated flour, so treated, is then fed into the flour stream at the rate of 2 lb per sack (280 lb). Unfortunately, this is not a very practicable method, and the miller finds the inclusion of chemicals far more satisfactory. The biscuit manufacturer is usually distrustful, however, and still prefers untreated flours. The fact remains that bleaching and stabilising can be brought about in this way, and the user should know how to detect these reagents and what action he can expect them to achieve.

Bleaches

Nitrogen peroxide: Nitrogen peroxide gas is produced by drawing air through an electric arc, and it is then passed into the flour stream. The degree of bleaching is regulated by the amount of air drawn in. Only very small quantities are used, 6 parts in one million being sufficient; flour that has been over treated in this way becomes over fermented very quickly.

Benzoyl peroxide: Also known as 'Novadelox'. Added as a powder at the rate of ¼-¾ oz per sack. There is no marked action until after 2-3 days.

Stabilisers

Persulphates—ammonium, potassium and sodium: These all have a very marked toughening effect on the gluten during

FLOUR AND CEREAL PRODUCTS 15

fermentation and this will lead to difficulties in handling and sheeting fermented doughs. Only ¼-½ oz per sack is necessary.

Phosphates—Acid calcium, sodium and potassium: Also have a toughening effect on gluten. Soft glutens are stabilised considerably. Phosphates are also yeast foods, so more vigorous fermentation will result if these are present. Acid calcium phosphate (the usual chemical) is used at the rate of 1 lb per sack.

Sulphates—Sodium and magnesium: These and calcium and ammonium sulphates have a similar action to the phosphates and in addition retard diastatic activity. 8 oz of sulphates are used per sack of flour.

Potassium bromate: Probably the most important. It has a remarkable astringent effect on gluten, only 0·1 oz is used per sack. It also has a stimulating effect on yeast activity.

N.B. These stabilisers would only be used to strengthen a flour which does not come up to the required standard for fermented doughs. There is no need to strengthen the weak flours for soft doughs, in fact, sometimes the reverse may be necessary.

Combined bleachers and stabilisers

The preceding chemicals perform either one or the other function, but some chemicals have a combined effect of bleaching and stabilising.

Chlorine: Chlorine gas is used at the rate of ½-2 oz per sack and not only oxidises the colouring matter, but also stabilises the gluten by increasing the acidity of the flour.

Chlorine dioxide: Also known as 'Dyox'. This is an extremely powerful bleacher, only ⅓-4 g per sack are necessary, not only to bleach the carotene, but also to have some effect on the colouring matter in the bran. Because of this it is very useful to use on low grade flours, and owing to the small quantities involved it is very difficult to detect.

Fungal enzymes: These are enzymic preparations produced from mould, yeast, or bacterial sources, usually in powder form. Those prepared from moulds are the most widely used in baked goods, as their action is readily destroyed during the baking process by the rise in temperature. Cereal enzymes are less readily destroyed during baking and consequently function for a longer period of the baking process. Bacterial enzymes may even survive the baking process (depending upon source) and could continue their activity in the baked article. Fungal enzymes may be added to the flour by either the miller or the manufacturer to remedy certain deficiencies in the flour. In biscuit making a proteolytic fungal enzyme can be used to assist in the breakdown of gluten in hard doughs where an exceptional degree of extensibility is required. The main drawback to the use of fungal enzymes is that their action is progressive. In other words, the breakdown continues while the dough stands. If this breakdown is achieved by mechanical or chemical means, there is no further action after the required degree is obtained.

Melloene treatment: Sulphur dioxide and steam are drawn into the flour stream, causing a marked softening of the gluten (increase in extensibility). This action is similar to the use of sodium m-bisulphite in hard doughs, but in this manner only half the available sulphur dioxide is necessary for the same results.

It is also extremely difficult to detect its presence. The main disadvantage is that the treatment wears off during storage until the flour reverts to almost its original state. Treated flour should therefore be used within a month of milling. Biscuit flour is permitted legally to contain up to 200 parts per million of sulphur dioxide.

Legal additives

In addition to these additives that the miller may or may not use, there is a short list which he must add to all flours except wholemeal. These additives are intended to cover deficiencies in the nation's diet and must be added according to the instructions of an Act of Parliament.

(1) Creata preparata (calcium carbonate): 11-17 oz per sack (usually 14 oz are used: this quantity is also 235-390 mg per 100 g)
(2) Vitamin B_1 : not less than 0·24 mg per 100 g
(3) Nicotinic acid: 1·6 mg per 100 g
(4) Iron: 1·65 mg per 100 g

FLOUR CONTAMINATION AND INFESTATION

Insects

Wheat is subject to several forms of disease, such as rust, mildew, smut, and ergot, but as these are all readily detected they should not be conveyed to the flour. Flour, then, is only normally open to attacks from insect life and airborne spores. (In breadmaking, grain-borne spores cause disease in bread, but these do not apply in biscuit manufacture.) Insect infestation is probably the most prevalent form of trouble, and this may occur by self-introduction to the flour, or by eggs present in the wheat surviving the milling process and hatching out in the flour during storage. Other sources of egg introduction are from poorly cleaned machines, bins, flour bags, and silos.

The commonest insect pest is the *flour moth,* which resembles the clothing moth in appearance, although it is generally a little larger. It is a very poor flier, and usually will take to the air only when disturbed. After mating, the female will lay up to 200 eggs in flour or other suitable foodstuffs. In favourable conditions the eggs hatch in a week or two, producing pinkish white caterpillars which move around in the flour, spinning a fine thread behind them. These threads announce their presence quite unmistakably. After about 10 weeks the larvae spin themselves cocoons in a secluded place and pupate. Three weeks later they emerge as moths, ready to start the life cycle once again. The moth in its various stages of metamorphosis is easily removed from flour by sieving; and if carefully sieved they appear to have no adverse effect on the flour, although in very bad cases the flour flavour is spoiled.

Rice or flour weevil are perhaps the next in order of importance as a flour pest. This is a small reddish-brown insect, about $^3/_{16}$ in (5 mm) long with a hard boring proboscis. The proboscis is used for boring holes in the grain, into which the female deposits her eggs. When the larvae hatch, they tunnel through the grains until they pupate. On completion of the metamorphosis, the adult eats its way from the grain. If these insects occur in flour, it is usually because the eggs were present in the wheat, but once established in flour, they will continue to reproduce if not dealt with effectively. They appear to have little effect on flour and can be removed by sieving.

Cockroaches are rather large nocturnal beetles, which breed readily in bakeries. They do not usually infest flour, but they do taint flour and other foodstuffs with their droppings, and are regarded as disease-carrying insects.

The *flour mite* on the other hand, renders flour unfit for human consumption by the high bacterial content of its excreta. It is an extremely small pest, only about $\frac{1}{32}$ in (1 mm) long. It is whitish in colour, with eight pinkish legs. The flour acquires a musty smell and a specky appearance when infested. If a pile of flour is smoothly compressed, the mites will struggle to the surface, causing little mounds.

Spores

Only in extreme cases of long storage, under moist conditions, will mould spores develop to any degree, giving rise to a musty smell. The common mould producing this condition is *mucor mucedo;* a white mould, which infiltrates throughout the flour in threads.

Flour storage: The ideal condition for flour storage is a well-ventilated, dry room, maintained at an even temperature of 16°C (60°F). Flour bags should not stand on the floor, but should have air circulating under as well as around them. When standing vertically, they should be no more than three high, and no more than five when horizontal. Efficient stock rotation must be maintained, and all ledges, corners, and the floor, must be kept free of dust and dirt.

CEREAL PRODUCTS

Oats: Oats are grown in cool temperate climates under conditions less favourable than wheat. They have a high

FLOUR AND CEREAL PRODUCTS 19

nutritive value, owing to a high proportion of proteins and fats. The protein is not gluten forming, and consequently, a dough cannot be made. The fat content is liable to turn rancid during storage, unless stabilised during treatment. Oats are available as meal in three or four grades, and also as rolled oats in a pre-cooked form. They can be used in biscuit production to bring about interesting changes in flavours and textures.

Maize: Maize, also known as Indian corn, grows in warm, temperate, and sub-tropical climates. It has a high starch content, the protein content is not gluten forming. Its main use in biscuit making is in the form of cornflour, which is milled from the endosperm of maize and is very nearly pure starch. Because of its high starch content, it can be used to dilute or weaken, flour that is too strong. Maize is widely used in the production of glucose.

Rice: Rice grows in eastern tropical climates. It has low fat and protein contents (non-gluten forming) and a high starch content. The milled grain is available as ground rice and rice flour, both of which have little application in biscuit making, except as a flour dilutant and to close the open texture of a biscuit at the same time as changing the flavour and eating texture.

Rye: Rye is mainly an East European grain, and apart from wheat, is the only one containing gluten-forming proteins. Its main uses in biscuit making are in speciality and 'health'-type biscuits. It is available as flour, various grades of meal, and in rough, broken pieces.

Barley: For growth, barley requires similar conditions to wheat. It is hardly used, if at all, in its own right, but is the main source for the production of malt and malt products.

Malt production: Malt is produced by a process of cleaning, softening, and germinating barley (and sometimes wheat) until roots and shoots are well developed on the grain. Growth is then arrested and the grain is dried in a kiln at 38°C (100°F) for 24 hr. The temperature may then be increased according to the type of malt required. The roots or shoots are brushed off, and the resulting treated grain is known as malt.

During the growing period, the diastatic enzymes become very active, they will remain active unless destroyed by heat or alkali. The action of diastase is to break down starch into dextrin and maltose (malt sugar). The sugar is used by the yeast in fermented doughs, thereby hastening fermentation and gluten modification.

If the kilning temperature remains low, then the enzymes remain very active; as the temperature is increased, the enzymes become decreasingly active until they are completely destroyed. Correspondingly, as the temperature increases more and more, sugar becomes caramelised and the malt becomes darker in colour. There is also a marked increase in flavour. Colour is a good guide to the diastatic activity and flavour of a malt product. A light coloured malt will be very active and will have little flavour. A dark malt will be inactive, but will have a strong flavour.

Malt products

Malt flour: Malt flour is produced by milling the malt and sieving. Unfortunately, the bran is very brittle, so it breaks up finely and is difficult to remove. The bran particles spoil the colour of the malt flour and it is difficult to produce a pale or white malt flour. Malt flour should be kept in an air-tight container, otherwise it loses flavour and, because of its hygroscopic nature, goes hard.

Malt extract: Malt extract is produced by preparing an extract in water of crushed malt. The liquor is filtered and evaporated in vacuum pans to the required consistency. Low temperature evaporation in vacuum pans is necessary to avoid altering the diastatic activity determined during the kilning operation.

Dried malt extract: Dried malt extract is produced by drying malt extract in a vacuum oven. The residue is milled very finely to a crystalline powder that is extremely hygroscopic. This type is convenient for weighing and dissolving.

Classification of malt products Malt products are classified, according to their diastatic activity, in degrees Lintner. This is a measure of how much starch can be converted by the group-enzyme diastase into dextrins and maltose.

Uses of malt products: The main use of malt products is for flavour, either in its own right, or to enhance other flavours. In fermented doughs, malt products act as yeast food, modify gluten, and improve flavour. A diastatic malt will also supply sugar from starch. In most biscuits, malt products with a good flavour are used, and diastatic activity is of minor importance.

Arrowroot and soya flour

Although neither of these are cereal products, it is convenient to deal with them at this juncture.

Arrowroot: Arrowroot is almost pure starch, obtained from the underground stems of a tropical plant. Because of the high starch content it may be used to weaken flours of too strong a nature, and is used in Arrowroot biscuits.

Soya flour: Soya is an extremely nutritious bean of northern Chinese origin which is now grown widely, particularly in the USA. It is available as flour in three forms: unprocessed and processed, full-fat soya flour, and as low-fat processed soya flour.

The unprocessed form is rich in fat and the enzymes are active, these are assorted and complex in their action, but those which are diastatic in nature can be of importance during fermentation as a supplier of yeast food. The processed forms of soya flour are inactive enzymically, and the characteristic bitter flavour of soya is rendered bland. Each type has a very high nutritive value, and the low fat type has a particularly high protein content (the fat content is extracted for use in margarine manufacture).

The main function of soya flour, when used in doughs, is as an emulsifying agent, owing to the presence of lecithin. This emulsifying action helps to produce a more homogeneous dough, which in turn should assist in biscuit-piece formation and sheeting, it should also help to prevent toughening and its attendant evils. Soya has been widely used as an egg and milk substitute. Its inclusion in doughs will result in an increase of biscuit colour and bloom.

CHAPTER 2

Fats and oils

THE terms fat and oil are only loose descriptive words, bearing little or no technical significance, except to describe the physical state of a fat. If a fatty substance is usually solid at a temperature of 15°C (60°F), it is normally termed a fat; if liquid at that temperature, then it is usually called an oil. Fats and oils are distributed relatively freely throughout nature and can be classified as mineral, animal, and vegetable.

Mineral oils cannot be digested by the human body, and also have a harmful effect in that they absorb vitamins from other foodstuffs. Members of this group are known as hydrocarbons and are only of interest and use as fuels and lubricants. It is illegal to incorporate these oils in foodstuffs, even those considered suitable for internal use medicinally.

Animal fats, with the exception of butter, are usually situated as a protective layer both just under the skin and around delicate and vital organs such as the kidneys. Butter occurs naturally as the fat content of milk. Animal fats are generally solid at 15°C (60°F).

Vegetable oils (usually liquid at 15°C (60°F)) occur in fruits, seeds, and nuts. These are probably the most important group in biscuit making, as the majority of manufacturers use oils of a vegetable origin, except where speciality fats are necessary or desirable.

Both animal and vegetable fats of this type are known as fixed oils, and should not be confused with the 'essential oils' which are volatile and highly aromatic. Fixed oils are greasy and will leave a permanent stain on paper, whereas the volatile essential oil will stain paper initially, but will gradually evaporate.

A further classification of fixed oils is whether they are 'drying' or 'non-drying' oils. Linseed oil is a drying oil and, as such, is unsuitable for biscuit manufacture; it is widely used in paint making because of its drying properties or capacity to form a skin or dry surface when in contact with air.

ANIMAL FATS

Butter

Butter is produced by separating the cream from cow's milk, allowing flavour to develop during a maturing period; the cream is then agitated or churned. This causes the fatty globules to coalesce and form granules of butter, the buttermilk or liquid is run off and the butter is well worked to a homogeneous mass, salt is added to improve flavour and keeping qualities, colour may also be added to improve or maintain a consistent colour.

The composition of butter varies according to source, but the fat content will be between 82% and 85%, the moisture content from 10% to 15%, the protein (known as curd or casein) content approximately 1·5%, and the salt content in the region of 2·0%. The fat content contains a small proportion of unstable volatile fats, these are responsible for flavour, but under unfavourable conditions of storage or with age, they break down and cause rancidity.

Butter should have a good, clean, fresh flavour and aroma; although flavour varies from source to source, it should always be free from cheesey or rancid taints; it should have a firm, even, and plastic texture, and good creaming or air retaining qualities. The colour and appearance should be fresh. Darkening and streakiness tend to develop with age. The salt and moisture content should not be excessive.

Butter can be used in most biscuits, but owing to its high cost, this is not an economic proposition. Its use, therefore, should be restricted to biscuits and goods which will make the most of the special flavour of butter and which will return the higher ingredient cost. For instance, shortbread relies on butter for flavour. Nothing from a bottle will satisfactorily take its place and, generally speaking, the public are prepared to pay for a good quality shortbread. There is very little to be gained by using butter in a biscuit of marked or strong flavour, as the butter flavour will be neutralised or masked. The use of butter can, however, be a sales point in a status conscious market, but its inclusion must be stressed and in accordance with the legislation upon this subject.

Refrigerated storage below 4·5°C (40°F) is ideal for butter and, if below freezing point, butter will keep for considerable periods. The shorter the storage period, the better will be the results as far as flavour and subsequent keeping qualities are concerned. For ease of handling, butter from a cold store should be allowed at least two days to attain room temperature; if this is not always possible, then the butter should be pulped and creamed before use. The flavour and texture of butter deteriorate if melted to oil, and this should be avoided.

Lard

Lard is the rendered fat from pigs, and the quality depends on the part of the body and method used in rendering. The highest quality lard is rendered in warm water not exceeding 32°C (122°F) from kidney fat, and is known as 'neutral lard No. 1'. The lowest quality is obtained under steam pressure from all the trimmings, and is known as 'prime steam lard'. Because of the low temperature used in rendering the high grade lards, there is a tendency to poorer keeping qualities. Where the low grade lards are subjected to high temperatures, they have better keeping qualities, owing to the destruction of enzymes.

Lard should be pure pigs' fat and have a pure white colour, granular texture, and a firm consistency. Low grades have poor colour and flavour. The flavour of lard is quite strong and is not suitable for sweet biscuits. Its use would be all right in savoury or unsweetened biscuits, although a blend may be necessary. Margarine manufacturers sometimes use lard in their products.

Beef fat

Beef fat is rendered from the fat tissues of beef cattle, and although it is used in the production of margarines and shortenings, it has no place in biscuit making in its own form.

Whale oil

The whale is one of the main sources of animal oil, which is obtained by rendering the thick layer of blubber by steam. At normal temperatures it is an oil with a strong odour, which is widely used in margarine manufacture after it has been deodorised and hardened.

VEGETABLE OILS

Vegetable oils are extracted from seeds and nuts by expression, or by solvents. In both processes the seeds are usually ground to a meal, and then the meal is subjected to heat and pressure in the mechanical expression process to sqeeze out the oils, leaving the solid material. A number of solvents may be used, such as light petroleum, carbon disulphide, carbon-tetrachloride, dichloroethylene or trichloroethylene, in the solvent process. Fats and oils are soluble in these solvents, and so the oil is extracted from the meal, the extract is filtered from the meal, and then the solvent is distilled or evaporated off, leaving the oil. The extracted oils then follow a process of refining.

The oils usually contain free acids which are neutralised with caustic soda, this action forms soap which is then washed away with water which is evaporated off in turn. Many of the oils have strong colours, and the next process bleaches the oil by the inclusion of bleaching earth which removes the colour by absorption. The oil is then filtered ready for deodorising. To remove the objectionable odours of the crude oils, steam is blown through the heated oil under vacuum, distilling off the odoriferous substances. The bland, colourless and odourless oil is now ready for processing or hardening.

Hardening is usually known as hydrogenation, as it is a process whereby hydrogen is added to an oil to convert it to a fat solid at normal temperatures. The reaction takes place under pressure, hydrogen gas is forced into the heated oil, agitation occurring throughout the process which may last from 2-6 hr. It is necessary for a catalyst, usually finely divided nickel, to be present during the process, and this must be filtered off on completion. The treatment causes a slight decomposition of the oil, and so the hydrogenated fat must be neutralised, bleached and deodorised before use. Apart from hardening the oil, hydrogenation brings about an improvement in colour and keeping qualities, whale oil, for instance, is unsuitable for edible purposes after the refining process, but hydrogenation achieves a remarkable improvement, making it perfectly suitable.

Coconut oil

Coconut oil is obtained from the dried white flesh of the fruit of the coconut palm, which grows in equatorial regions. The main sources are Ceylon, Indonesia, the Philippines, the West Indies, Equatorial Africa, and the South Pacific Islands. The refined oil is a white brittle fat with a fairly sharp melting point at 25°C (77°F) and passes rapidly from a hard solid to a liquid. By hydrogenation it is possible to produce a fat with a longer plastic range and a higher melting point of 32-34°C (90-93°F).

Coconut oil is particularly suited to biscuit cream fillings, owing to this property of setting quickly to a hard brittle fat. Hardened coconut oil will be necessary under normal summer conditions, but in winter the melting point of the un-hydrogenated oil should be high enough to prevent melting in the finished biscuit during storage. Coconut oil is quite unsatisfactory as a shortening in doughs because of the transition from solid to liquid state being so rapid and at a temperature close to operating temperatures. If the fat does not become oily in the mixing period it can very easily do so at the machining stage, resulting in trouble with sheeting and blocking up of the impressions on the rotary moulder.

Coconut oil can be used satisfactorily in a blend of fats for dough making and also in the flour/fat mixture used for layering in cream crackers and in puff doughs.

Palm oil

Palm oil and palm kernel oil are both the product of the oil palm, a tree indigenous to West Africa, but also grown in South-East Asia. Although produced from the same fruit, the oils are remarkably different. Palm oil is produced from the outer fleshy portion of the fruit, about the size of a small plum, which grows in clusters weighing about 40 lb. Palm kernel oil is produced from the nut inside the stone of the fruit.

The flesh of the palm quickly deteriorates, and so the oil has to be extracted in the growing areas, producing a yellow to red crude oil, which is processed in the country where it will be used. Palm kernels, however, keep well and are extracted and processed in the country of use.

Refined palm oil is a whitish yellow colour with a soft plastic nature over a long temperature range. Its slip melting point is not easy to ascertain, but should be between 35 and 38°C (96-101°F). The oil can be hydrogenated to give any particular melting point, but temperatures above 40°C (104°F) are more satisfactory than those below.

Palm oil can be used satisfactorily in all biscuit doughs. If the prevailing temperatures cause oiling, then a higher slip melting point will help overcome this difficulty. Although palm oil can be used as the only shortening in doughs, it is more usual to use a blend with palm oil as the base. It is also widely used in margarine manufacture.

Palm kernel oil

Refined palm kernel oil is very similar to coconut oil in appearance, hard and white, but not quite as brittle. Although the slip melting point of palm kernel oil is 28-29°C (82-84°F), slip points up to 48°C (118°F) can be achieved by hydrogenation. Hardened palm kernel oil is ideally suited to biscuit filling creams and it can be used wherever coconut oil would be used. Both are very stable fats with good keeping qualities.

It is also possible to separate from palm kernel oil speciality fats resembling cocoa butter in melting point and texture, but which are not so expensive.

Groundnut oil

The groundnut (also known as peanut, monkey nut, and arachis) is a seed grown in a pod underground. The plant orginates in South America, but is widely grown in tropical, sub-tropical, and Mediterranean-type climates. Refined groundnut oil is liquid at normal temperatures, but at lower temperatures produces a crystalline deposit. The oil is almost colourless and has a bland flavour. It can be hydrogenated to produce quite high melting points. It is widely used in margarine production and in salad and frying oils. In biscuit making it is frequently used in shortening blends, as it gives a long plastic range and improves the quality of the blend and consequently, the biscuit. It is not satisfactory to use in creams, even when hydrogenated, as it has a tough rather than a brittle texture.

There are other vegetable sources of oils, but none of them are as important to the biscuit manufacturer as those already mentioned. Olive oil is derived from both the flesh and the kernel of the olive, but is too expensive for normal biscuit making, and groundnut oil is used widely as a substitute. Maize, soya beans, cottonseed, and the cocoa bean, are all oil producing and may be used by the oil refiners in specialist fats and blended products.

Margarine

Margarines are manufactured by blending different types of oils to produce a fat of the required consistency, texture and flavour at a specified price. Usually, a soft oil such as groundnut oil, a semi-soft oil such as palm kernel oil, and a hydrogenated oil such as whale, are blended in the desired proportions while hot, and are then cooled. Just before achieving their setting point, a culture of milk, salt, emulsifier, colour and flavouring agents is added and vigorously mixed in. The moisture content must not exceed 16%. (Domestic margarine also has vitamin concentrates added at this stage.) The emulsion is then chilled in a very fine film on a refrigerated drum and is scraped off in fine flakes. The solid flakes are then kneaded to form a homogenous plastic product, similar to butter. Margarine competes very favourably with butter in texture, consistency, keeping qualities, constancy, and price, but not in flavour. Margarine flavour is developed by controlled souring, and often bacterial action, of the milk culture prior to adding to the blend, but in spite of a great deal of research and extravagant claims, no flavour has yet been produced which equals the natural flavour of butter.

Margarines are produced for cake making and for biscuit doughs. These latter usually have an improving effect on doughs for machining and sheeting, and enhance biscuit flavour. Suitable margarines are also manufactured for the production of puff doughs. These blends have a higher melting point and a tough plastic nature. In the production of puff biscuits the dough has to undergo a lot of machining and layering, and only a tough extensible margarine could withstand this treatment and still retain the layered structure. Unfortunately, in achieving this type of texture, the melting point of the margarine exceeds body temperature and consequently does not melt in

FATS AND OILS

the mouth, leaving a nasty waxy lining on the palate. However, it is possible to obtain puff pastry shortenings at rather high prices, which have a tough plastic nature and yet with melting points lower than body temperature. Other specialist fats are also available that are hard in nature, but since they are not tough and plastic, they require chilling, mincing, and chilling again, before distribution in the dough pieces.

PURPOSE OF SHORTENINGS IN BISCUIT DOUGHS

If flour is mixed with water, a dough forms which is tough and rubbery, and if stretched or sheeted, it will spring back to almost its original size and shape. This action is due to the formation of gluten. It is possible to modify the gluten and so make it more extensible by physical development; by fermentation; or by the inclusion of shortenings.

The action of fat in a dough is more one of preventing the gluten forming than actually modifying it. As gluten does not form until the flour is in contact with water and mixing action, the inclusion of fat tends to insulate the gluten-forming proteins from the water and consequently, a less tough dough results. The greater the amount of fat, the greater the insulating effect will be. Excessive mixing will break down the insulation and a tough dough will again result.

A secondary action of fat is to 'lubricate' the gluten which has formed, allowing it to slip into a new position when sheeted or formed into biscuit shapes, without the same desire to return to its original position.

THE CHEMICAL NATURE OF OILS AND FATS

The chemistry of fats and oils is extremely complex and deep, but superficially it is probably only necessary to know that they are organic compounds composed of carbon chains to which are combined hydrogen, and in a small proportion, oxygen. They are the result of a chemical reaction between glycerol (glycerine) and a fatty acid or fatty acids, forming a glyceride and water thus:

Glycerol + Fatty acid → Glyceride + Water
e.g. Glycerol + Stearic acid → Stearin + Water

One molecule of glycerol is capable of combining with three molecules of fatty acid, and this being the case, a tri-glyceride results which is a true fat or oil. If only one or two molecules of fatty acid combine, then a mono-glyceride or di-glyceride (respectively) forms. These are not fats, and owing to their chemical composition they are soluble in both water and in fats, which makes them important as emulsifiers, and as stabilisers in emulsions.

If a tri-glyceride forms where the glycerol has combined with molecules of the same fatty acid, then this is known as a simple glyceride. It is, however, more often the case in fats for glycerol to combine with two or three different fatty acids to produce a mixed glyceride.

As the melting point and nature of the fatty acid influences the melting point and nature of the fat, it follows that a fatty acid with a high melting point will produce a fat with a high melting point and vice versa. On the other hand, the melting point of a mixed glyceride will depend on the melting points of the fatty acids involved and the proportion of one to the other.

The hard fatty acids with high melting points, which are chemically complete and stable, are known as saturated fatty acids. The soft fatty acids with low melting points are unstable and are chemically incomplete. These are termed unsaturated fatty acids. The long carbon chain in the molecule contains what is known as a 'double bond'. This double bond is the unstable link in the chain which seeks to join up with hydrogen to become a stable saturated fatty acid. This action is basically what occurs during hydrogenation. By the addition of hydrogen, the soft unsaturated fatty acids become hard saturated fatty acids, resulting in the hardening of an oil, with an increase in the melting point temperature. Unfortunately, during storage of unsaturated fatty acids, there is no hydrogen available (or catalyst, heat, and pressure), so the fatty acid joins up with oxygen which is available in the atmosphere, resulting in oxidation and the effects of rancidity.

Double bonds may occur once, twice, or even three times in the carbon chain and consequently, the oil becomes more and more unstable, and also more readily hydrogenated.

Stearic, lauric, and palmitic acids are common saturated fatty acids that are hard and have a high melting point. Oleic, lauroleic, and palmitoleic acids are unsaturated fatty acids with one double bond (mono-unsaturated) and low melting points.

FATS AND OILS

Linoleic acid is a di-unsaturated acid and linolenic acid is a tri-unsaturated acid, both of which have melting points lower than 0°C. These examples are of the commoner fatty acids present in fats and oils, and some or most of them will be found in nearly all oils to a greater or lesser extent in conjunction with less common ones, each one influencing the nature of the fat in which it is a constituent part.

When a fat contains proportions of the di- and tri-unsaturated acids it tends to be rather unstable, but it is possible by 'selective' hydrogenation to hydrogenate these only and so improve their stability without increasing the melting point of the fat to any great extent. For instance, palm oil containing about 6% linoleic acid and traces of linolenic acid is prone to oxidation, but by a small amount of selective hydrogenation they both can be converted to mono-unsaturated acids with greater stability and little alteration of the melting point.

RANCIDITY

Rancidity is the name given to the state when a fat or oil breaks down and develops unpleasant odours and flavours. These are usually accompanied by an initial darkening of the fat and finally becoming bleached. The strong and characteristic odour and flavour make the fat unusable in foodstuffs. Rancidity occurs during storage by either enzymic action or by oxidation.

Enzymic Rancidity

(1) *Hydrolytic:* This occurs by the action of fat-splitting enzymes and the addition of water, producing a reversal of the reaction illustrated previously. The fat (glyceride), becomes combined with water to form glycerol and fatty acids.

(2) *Oxidising enzymes:* In this case, fat-splitting enzymes break down the saturated fatty acids of high molecular weight by oxidation, to form low molecular weight volatile compounds with the characteristic rancid odour.

In refined oil products there should be no danger of either of these enzymic rancidities occurring, as the high temperatures destroy any enzymes present and the oils are completely devoid of moisture (margarine contains water, but should not contain enzymes). However, in products such as butter and low-temperature rendered lard, enzymic rancidity is not only a possibility, but a fact which must be borne in mind when storing this type of product.

Rancidity by oxidation

This is the most common form of rancidity and occurs when the oxygen of the atmosphere combines with the unsaturated fatty acids to produce volatile fatty acids and other odoriferous compounds. Only fully hydrogenated fats are exempt from this form of rancidity, and as there are very few completely hydrogenated fats used in biscuit manufacture, it is as well to assume that all fats are subject to rancidity by oxidation.

Oxidation of fats will be accelerated by storage in warm conditions and by exposure to light (particularly direct sunlight and ultra-violet light). Contact of metals, such as copper (and cuprous metals), are powerful oxidising agents, and should be avoided.

Lecithin

Lecithin is a phosphatide, that is, it is closely allied to fats, yet contains phosphorous and nitrogen. It is distributed fairly widely in nature, particularly in foodstuffs rich in fat and protein. Commercial lecithin is obtained chiefly from soya beans and groundnuts, but other seeds and nuts yield lecithin during the fat extraction and refining process.

It is an important raw material as an emulsifying agent, particularly for water in oil emulsions. It is used extensively in blends of oils, whether they contain water or not, as it helps to maintain a homogeneous mixture and there is less possibility of separation if the different oils have different specific gravities. If lecithin is added to biscuit-filling creams, a noticeable increase in fluidity can be achieved. This can be very important if (1) a lower fat content is required in a cream, (2) the cream carries a proportion of ground-down biscuit waste or, (3) to maintain a constant sugar to fat ratio and standard handling methods during changing temperatures and conditions.

FATS AND OILS

The quantities necessary are very small, and if based on fat content, 0·5% will probably be the maximum required.

Lecithin may also be used in biscuit doughs to improve fat distribution and emulsification of the ingredient. The most suitable method of incorporation is for the lecithin to be added in with the fat blend.

CHAPTER 3

Sweetening agents

THE majority of sweetening agents are obtained from sugar in one form or another, and sugar (sucrose) is derived from two sources. The main source is sugar cane and the secondary source is sugar beet.

Sugar cane is a tropical plant, similar to bamboo in appearance, grown mainly in the Caribbean Sea area. The plant contains from 20% to 25% sugar which is extracted by crushing and soaking. The extract is filtered, and then clarified by treatment with heat and lime. This juice is then concentrated in vacuum pans and allowed to crystallise. Excess moisture is spun off in centrifuges to yield raw sugar which is also known as Muscovado. The raw sugars vary considerably in crystal size and also in colour, which ranges from pale yellow to dark brown. They are usually known by their source of origin thus: Demerara, Trinidad, or Barbados, all of which have differences of appearance and nature. The residual syrups that are unsuitable for crystallising are fermented for distillation as rum.

Sugar beet is a root crop which grows in temperate climates, mainly Europe, yielding 15% to 20% sugar on extraction. The beet are shredded and soaked in water, clarified, evaporated, and crystallised, ready for refining.

Cane and beet sugar are very similar in the raw state, and after following the same refining process, are identical as first grade sugars. Cane sugar tends to be slightly better in appearance in the lower grades.

SUGAR REFINING

The raw sugar is first sprayed to remove excess syrups and is then dissolved and filtered. The sugar solution has milk of lime added and carbon dioxide is pumped through resulting in calcium carbonate being precipitated which clarifies the solution of many of the impurities such as waxes, gums, and tannins. The resulting yellowish syrup is then passed through cylinders of activated charcoal which absorbs the colour.

The clear pure solution is then evaporated in vacuum pans to avoid the dangers of caramelisation and discoloration, and when

SWEETENING AGENTS 35

the correct concentration is reached, the solution is crystallised by the introduction of crystals of sugar. The crystallised mass still contains syrup, which is removed by centrifuges. The crystals are sprayed, to remove all the syrup, and recentrifuged. The reclaimed syrup is filtered again through charcoal, reboiled, and crystallised again. This process yields second grade sugars and, if repeated, third grade, but after that the reclaimed syrup is discoloured, but is used in golden-syrups and low grade sugars, such as 'fourths', or 'pieces'.

The refined sugar contains a small amount of moisture which has to be dried out, and the crystals are then graded for size to give three grades: granulated sugar, castor sugar, and sugar dust.

SUGAR GRADES

The sugar grain size is dependent upon the syrup concentration and speed of crystallisation in the vacuum pans. Small crystals can be increased in size by the addition of more syrup, but the reverse is not possible except by reboiling and recrystallising. It is possible directly from boiling to produce granulated sugar in three grades: coarse, medium, and fine; castor sugar, which is the finest grade of sugar crystals; and coffee crystals, which are quite large crystals grown by feeding with more syrup in the vacuum pans.

Lump or cube sugar is produced from setting sugar in flat blocks and then cutting into cubes. Nib sugar, is broken pieces of sugar produced during the cutting of cube sugar. Modern centrifugal methods of preparing cube sugar may ultimately lead to difficulties in obtaining nib sugar.

Pulverised sugar is produced by milling granulated sugar and sieving into two grades: pulverised castor—a very fine crystal with very little dust; and pulverised icing which is mainly icing sugar with some fine crystals remaining.

The finest grade of all is icing sugar which is milled extremely finely to a powder and should contain no crystals.

Pulverised and icing sugars will vary considerably in particle size from factory to factory when milled on the premises (the normal practice), but the previous notes are a guide to grades available from the sugar refiners.

Lower grade sugars are also available as granulated or soft sugars, but the colour deteriorates as the grade gets lower until the fourth and subsequent grades, or pieces, are usually a pale to dark brown in colour with small, soft, moist crystals. These lower grades are frequently used in the production of refiner's syrups.

SUGAR PRODUCTS

Golden syrup

Golden syrup is made by the refiner from low grade sugars and uncrystallised syrups. It is a mixture of sucrose and invert sugar in solution with a small proportion of gums, acid, and mineral salts. The colour is partly natural and partly added to a standard colour. It is usually available in two grades, the lighter coloured being purer, but having less flavour, while the darker, with more impurities present, is made from lower grade sugars and syrups and has a stronger flavour. This darker syrup is sometimes known as 'manufacturer's syrup'.

Treacle or molasses

This is the dark coloured uncrystallised syrup which remains after the production of raw sugar. It consists of a solution of sucrose, invert sugar, and various 'impurities', which impart the colour and flavour.

Invert sugar or syrup

Invert sugar is a clear syrup produced by the action of an acid on a sucrose solution yielding equal parts of glucose and fructose. (These simple sugars are also known as dextrose and laevulose respectively.) One important factor of invert sugar is its great affinity for water.

SWEETENING AGENTS OTHER THAN SUGAR PRODUCTS

Honey

Honey is a natural product made by bees from the nectar of flowers. It is composed mainly of invert sugar and a small quantity of sucrose in solution. The distinctive flavours of honey depend upon the essential oils present in the pollen of the flowers from which nectar is gathered. The excellent colour and fine flavours of English honey are derived from three major sources: clover, sycamore, and heather, but these are expensive for commercial use. Imported honeys are usually cheaper, darker, and much stronger in flavour, and are therefore more suited for biscuit making. Care is necessary in using imported honeys, as there is sometimes a chance of very strong flavoured pollens being present which will spoil the honey (for example eucalyptus in Australian honey).

Glucose (commercial) also known as corn syrup

Commercial glucose is produced by acid or enzymic hydrolysis of maize, potato or wheat starch. As this conversion is not a direct step, there are always intermediary products present. In addition to glucose (dextrose) in solution, there are also dextrins and maltose. Glucose is a clear, thick, viscous fluid, only about half as sweet as sucrose. Glucose is readily fermented by yeast. During baking it readily caramelises, giving a good colour to the face of the biscuit, and it also assists in soft dough formation.

Recent advances in glucose manufacturing techniques have resulted in the ability to produce a wide range of specifications. The monosaccharide content of glucose is largely responsible for its sweetness, and browning properties during baking. By controlling the composition of glucose it is possible to control the degree of sweetness and the degree of browning. Similarly, viscosity and the hygroscopic nature of the glucose can be significantly affected. A high maltose glucose syrup (with a low sweetness) when spray dried, could be used as a filler in savoury biscuit creams without greatly affecting the sweetness of the cream.

Malt extract

Malt extract is very poor as a sweetening agent, but consists of approximately 50% malt sugar (maltose), the rest being water, dextrins, and a small proportion of protein. Its main use is for flavour, but can also be used in fermented doughs to assist gluten modification and as a yeast food.

Malt syrup

Whereas malt extract is generally prepared from malted barley, malt syrup is produced from barley and maize starch and is consequently sweeter. It is used for flavour and as a yeast food.

Maple syrup

Maple syrup is a syrup with a characteristic flavour obtained from the maple tree by tapping. The flavour is only light, so should only be used to its best advantage.

CHEMICAL NATURE OF SUGARS

All sugars are members of the carbohydrate group of foodstuffs. Carbohydrates are organic compounds composed of carbon, hydrogen, and oxygen. The hydrogen and oxygen are always in the same ratio of 2:1, as in water. Carbohydrates are distributed very widely throughout nature, particularly in plant life, and are an extremely important souce of energy in our diet.

The carbohydrate group commences with the simple sugars such as glucose and fructose, known as monosaccharides. The more complex sugars, such as sucrose, maltose, and lactose (milk sugar), are known as di-saccharides. Other members of the carbohydrate group are no longer classed as sugars. They are complex and variable with the general description of poly-saccharides and include dextrin, starch, and cellulose. Plants build up carbohydrates by absorbing carbon dioxide from the atmosphere and moisture from the earth, in the presence of chlorophyll and sunlight. This process is called photosynthesis. Oxygen is released by the plant. Breakdown of the complex to the simple forms can be achieved by a process of hydrolysis.

Each stage is accompanied by the combining of water and a splitting of the molecule by the action of either an enzyme or an acid.

Monosaccharides have a common chemical formula ($C_6H_{12}O_6$), but are dissimilar in their properties. Glucose is a fairly sweet simple sugar occurring naturally in grapes. It will crystallise from a concentrated solution, and rotates polarised light to the right, hence its other name 'dextrose'. It is fermented by yeast to produce alcohol and carbon dioxide.

Fructose occurs in fruits and honey, but usually in conjunction with glucose as invert sugar. It is the sweetest of all the sugars and is fermented to alcohol and carbon dioxide, but does not readily crystallise. Polarised light is rotated to the left (laevulose). As invert sugar contains the dextro-rotatory glucose and the laevo-rotatory fructose, the rotation is partially balanced, but still remains laevo. Invert sugar forms when sucrose (dextro-rotatory) is hydrolysed, and as the rotation is reversed, the sugar is said to be 'inverted'.

The formula of the disaccharides ($C_{12}H_{22}O_{11}$) is also common to the group, but the members have different properties.

Sucrose is the commonest of the sugars and the most important to biscuit manufacturers. It is built up from one molecule of glucose with one molecule of fructose, minus one molecule of water. It is broken down by the enzyme invertase during fermentation to form invert sugar.

Maltose is not a very sweet sugar. It occurs in malt products, is dextro-rotatory and on hydrolysis by acid or maltase, yields 2 molecules of glucose. It is fermented in two stages by yeast.

Lactose is the sugar found in milk. It is only slightly sweet. On hydrolysis it yields glucose and galactose, but is not fermented by yeast.

EFFECTS OF VARIOUS SUGARS ON BAKED PRODUCTS

The initial purpose of sugar in baked goods is to sweeten the products, but by varying the grades and types of sugars, various other effects can be achieved.

As the sugar content increases, the biscuit becomes harder, until a hard brittle biscuit with 'snap' is obtained. This effect can be lessened by other treatments such as an increase in shortening, or accentuated by slower baking and so achieving a greater degree of caramelisation. In addition to the 'bite' altering, the texture will become closer and the biscuit will spread in area.

The degree of spread is largely dependent upon the effect of softening or tenderising which sugar in solution has on the gluten. As the gluten becomes softer, the biscuit flows more easily, and the more sugar that dissolves, the more the gluten is softened. The sugar that is already in solution, e.g. golden syrup and glucose, or that which dissolves most easily, has the greater effect on the gluten and consequently, on the spread of the biscuit. As most biscuit doughs have a relatively low water content, the sugars that will dissolve most readily are those with the finest particle size. Thus pulverised icing will go into solution more easily than castor; and castor more readily than granulated.

However, if biscuit spread is prevented by less sugar being in solution through using a coarser sugar, the texture will also be opened up. The texture of a biscuit can be controlled by aeration to counter this, but if the crystal size becomes too great and too concentrated, it will show up on the biscuit face and base.

In the case of a flow type biscuit which contains sufficient sugar to spread considerably, and the baking is suitable to induce caramelisation and 'snap', then the addition of a coarse grained sugar will break up the surface of the biscuit into irregular fissures known as 'crack'. These fissures will vary in width according to the sugar grain size—the coarser the sugar the coarser will be the 'crack'.

Although white sugars influence size, bite, texture, and sweetness, they do not alter the flavour of the biscuits. The lower grade sugars and syrups, and the raw sugars, are very important, as they influence flavour considerably. It must be borne in mind, however, that the dark colour of the sugar can be detrimental to plain biscuits; on the other hand it can contribute to the colour of some biscuits.

TYPES OF SUGARS FOR OTHER PURPOSES

The main use of sugar except for baking is in the preparation of filling creams and icings. In these instances a fine icing sugar should be used, otherwise there is a tendency for them to be 'gritty' when eaten.

Whenever sugar is completely dissolved, as in marshmallow syrup (or even if used in fermented doughs), granulated sugar should be used, as this entails no extra labour costs. Different effects can be obtained by sprinkling various grades of sugar on biscuits before baking. It is general practice to use fine or medium granulated sugar. An attractive finish can be achieved by first washing the biscuit (with water, or milk, or a mixture of both) and then sprinkling with castor sugar. During the baking period some of the sugar is dissolved and a mosaic effect results with no loose sugar on the biscuit face.

Routine tests

Apart from checking the colour of the sugar, which can vary, especially if using a 'manufacturing' grade, the main concern is a check on particle size. This is carried out by sieving a standard weight of the sugar sample for a standard time through a nest of graded sieves. Variations in particle size will be shown up, and if the sugars are milled on the premises, adjustments can be made to rectify the fault.

CHAPTER 4

Aerating agents

THE basic ingredients for biscuits have now been dealt with. To increase the palatability and to improve texture, bite, and appearance, it is necessary to achieve some form of aeration. Methods of aeration can be classified broadly into three groups, namely: mechanical, biological, and chemical.

MECHANICAL AERATION

Mechanical aeration is achieved without the use of ingredients, but by the method of handling the product. Beating and whisking are the main methods, and in these air is introduced to the product and is then broken down into very fine bubbles by continued division caused by the blades of a beater or whisk. There must be some ingredient in the product, however, that will hold these air bubbles and not allow them to escape. This is usually brought about by a protein substance such as egg, albumen, or gelatine. Examples of this method are the production of marshmallow and icing.

Another method of mechanical aeration is in the use of layers of dough with an insulating material in between. Examples of this are the production of crackers and puff biscuits. In both these cases a dough is used with well-developed gluten present. The dough is sheeted and a layer of fat, or fat or flour, is spread on the dough. The dough is then folded over and sheeted, and the process is repeated until a system of thin layers of dough are alternated with thin layers of fatty material. When the biscuit enters the oven the water in the dough layers is converted to steam and expands, lifting the layers as it does so. The insulated fatty layers prevent a great deal of the steam escaping initially. With correct baking conditions as the biscuit becomes fully expanded, the gluten begins to coagulate and the starch to gelatinise, thus forming a rigid structure for the biscuit to retain its shape when it cools after baking.

BIOLOGICAL AERATION

To bring about aeration by biological means, yeast is used. Yeast is a microscopic, unicellular organism, that relies mainly upon sugars in solution for life and reproduction. It is capable of breaking down sucrose and maltose into monosaccharides, and glucose and fructose into alcohol and carbon dioxide. When compressed yeast is added to a dough, the correct conditions for life and reproduction are available; the yeast feeds upon the sugars and produces carbon dioxide, which in turn aerates the dough. However, in biscuit production fermentation is mainly used to bring about gluten modification and flavour development; the aeration that is achieved is of very minor importance, as the time between the dough passing through the last gauge roller until it reaches the oven is insufficient for any measurable amount of aeration. Small quantities of carbon dioxide will still be present in the dough and some will be produced rapidly until the oven heat destroys the yeast and this will expand in the oven causing some aeration, but the greater part of the aeration is caused by mechanical means already described.

CHEMICAL AERATION

By far the most widely practised and important method of aeration is the use of chemical substances capable of evolving a gas, but which do not leave unpleasant flavours or odours, nor have harmful effects.

Ammonium bicarbonate

Ammonium bicarbonate is an entirely volatile salt (hence its common name 'Vol') which, when heated, liberates carbon dioxide, ammonia gas, and water. Ammonium carbonate acts similarly, but is not as stable as the bicarbonate and tends to decompose during storage. Both are excellent aerators, but must be used with discretion. Excess can completely break down the structure of the biscuit, and it is important that the ammonia gas is eventually released from the biscuit because of its very strong pungent flavour and odour.

Sodium bicarbonate

Sodium bicarbonate is the most important of the aerating chemicals and is used very widely. When heated, the carbon dioxide and water are released, leaving sodium carbonate as the residual salt. Sodium carbonate is known also as common washing soda, and as such has an unpleasant flavour and can react with fats (particularly if developing rancidity) to cause soapy tastes. Sodium carbonate has a marked softening action on gluten, causing spread, and it also darkens the crumb. In goods such as gingernuts, these properties can be used to good advantage and the poor flavour is masked by the ginger.

If sodium bicarbonate is heated, only half the total amount of carbon dioxide is released. If an acid is used to react with the alkaline salt, all the carbon dioxide is released and there is no noticeable action on the gluten or on the crumb colour. In biscuit making it is the usual practice to use an acid ingredient to react with part of the sodium bicarbonate, but it is rarely the practice to neutralise the alkali completely. In this way there is a considerable yield of carbon dioxide for aeration, as well as a smaller quantity of sodium carbonate being left to soften the gluten and permit the biscuit to spread a little. The flavour and discoloration will be minimised to such an extent as to be unnoticeable.

To be effective, aeration is desired when the biscuit enters the oven, otherwise the gas is lost during mixing and before machining, and since the acid/alkali reaction does not take place until both substances are in solution, one of the chemicals should not be soluble except in warm or hot water. Sodium bicarbonate, soluble in cold water, is the alkaline chemical (*N.B.* Sodium bicarbonate is actually an 'acid salt' but it reacts as an alkali.) It follows, then, that a suitable acid ingredient should be used which does not dissolve in cold water, but does so when the water is warmed by the heat of the oven. There are several acids available and in use, but few can be considered perfect.

Tartaric acid and citric acid

There appears to be no reason for using tartaric acid in preference to citric acid, or vice versa, in biscuit making, in doughs, or filling creams, except from a point of view of cost, and as the prices fluctuate considerably it is better policy to alternate with the market. As aerating ingredients, both are readily soluble in cold water and the reaction occurs immediately mixing commences, leaving only the unreacted sodium bicarbonate to produce aeration in the oven.

AERATING AGENTS

Cream of tartar

Cream of tartar has largely been replaced by cream powder (see p. 46). It is fairly soluble in cold water so the reaction commences immediately, but will generally reserve some aeration for the oven. 2¼ lb cream of tartar will completely neutralise 1 lb sodium bicarbonate (2·25 : 1). It is possible to use 'protected' cream of tartar, but this is rather expensive. The cream of tartar is coated with a layer of glycerol mono-stearate which is insoluble in cold water, but dissolves in hot water, reserving all the aeration for the oven. 6 lb protected cream of tartar are necessary to neutralise 1 lb sodium bicarbonate (6 : 1).

Acid calcium phosphate (A.C.P.)

Acid calcium phosphate is also known as monocalcium phosphate. It is fairly soluble in cold water, but for goods reaching the oven soon after mixing it reserves a good proportion of the reaction for the baking process. There is a slight toughening effect on the gluten. Acid calcium phosphate is generally regarded as being adequate as the acid ingredient with sodium bicarbonate when used in conjunction with ammonium bicarbonate for biscuit aeration at a very reasonable cost. 1 lb 4 oz acid calcium phosphate is required for complete neutralisation of 1 lb sodium bicarbonate (1·25 : 1).

Sodium acid pyro-phosphate

Although sodium acid pyro-phosphate is slightly soluble in cold water its solubility, and consequently, the rate of reaction, increases with the rise in temperature. It has a softening effect on gluten. The main disadvantage of sodium acid pyrophosphate as an acid ingredient for aeration is that the residual salt, sodium pyrophosphate, causes a burning sensation at the back of the mouth. Fortunately, the amount used in biscuit making should be low enough for this characteristic not to be obvious. 1 lb 6 oz sodium acid pyrophosphate is necessary to neutralise 1 lb sodium bicarbonate (1·4 : 1).

Monosodium phosphate (inert)

The normal monosodium phosphate is readily soluble in cold water, but under a patent process of manufacture it remains inert in the cold state, but dissolves when heated. The residual salt, sodium phosphate, has only a mild salty flavour. 1 lb 7 oz monosodium phosphate neutralises 1 lb sodium bicarbonate (1·45 : 1).

Cream powders

There are numerous proprietary acid ingredients manufactured by the chemical industry which are generally referred to as 'cream' powders. The name refers to the fact that they are designed to replace, or improve upon, cream of tartar. They are usually prepared from the acid ingredients previously mentioned, either on their own, or in combination. Cream powders are normally diluted with an inert substance (or filler) such as rice flour, so that 2 lb cream powder neutralises 1 lb sodium bicarbonate (2 : 1).

Glucono-delta-lactone (G.D.L.)

Glucono-delta-lactone is not itself an acid, but when in solution it gradually converts to gluconic acid which then reacts with the sodium bicarbonate to release carbon dioxide for aeration. The conversion occurs only slowly at room temperature, but increases as the temperature rises, consequently the aerating effect is reserved mainly for the baking process. 2 lb glucono-delta-lactone neutralises 1 lb sodium bicarbonate and no after taste remains. A protected form of glucono-delta-lactone is available with a coating of calcium stearate.

Although aeration is intended to open the texture of the biscuit, and in so doing increase the palability and improve the bite and appearance, the acid and alkaline ingredients involved have a secondary effect on the dough. They affect the natural acidity of the dough. Flour is naturally acid in nature; sugar is inclined to the acid; the fat should be neutral; the water will vary from district to district—but on balance the mixed dough would be of an acid nature. If used in the correct proportions the aerating chemicals would neutralise each other and would

AERATING AGENTS

have no effect on the acidity of the dough, but normally the aerating ingredients are not balanced. The alkaline sodium bicarbonate is usually in excess and therefore has a neutralising effect on the natural acid and, depending upon the amount of sodium bicarbonate in excess, the dough may remain acid, or become neutral or even become alkaline. The degree of acidity, or alkalinity, is known as the 'hydrogen ion concentration' and is measured on a scale which gives figures known as pH values.

HYDROGEN ION CONCENTRATION

When substances such as sugar are dissolved in water they merely become dispersed in solution and are neither acid nor alkaline, but are neutral. Other substances, however, when dissolved, split into particles, known as ions, carrying electrical charges. These substances are acids, bases (alkalis are bases), and salts. The acids split up and form hydrogen ions with negative electrical charges, and the salts may be either acid or neutral (alkaline 'salts' are bases).

Pure water tends to form ions, but as the hydrogen ion is positive and the hydroxyl ion negative, and they are in equal numbers, they cancel each other out and water is said to be neutral. The neutral point on the pH scale is 7·0. Figures greater than 7·0 are alkaline, for example: 7·2 is very mildly alkaline; 10·0 is a considerably stronger alkali; below 7·0 denotes acidity. The scale covers 14 divisions. Each decrease of a whole number denotes a tenfold increase in hydrogen ions.

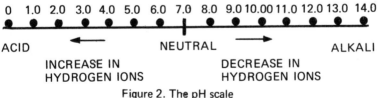

Figure 2. The pH scale

The pH of foodstuffs will generally be between pH 5·0 and pH 8·0; and most biscuits about pH 6·0.

DETERMINATION OF pH VALUES

(a) Colorimetric: Certain substances have the ability to change colour at a definite point on the pH scale. These substances are known as 'indicators'. For fairly accurate results a clear colourless solution or extract is necessary to which is added a solution of the indicator. Different indicators give colour changes over various pH values and so it is necessary to use an indicator applicable to the range to be tested. Universal indicators are prepared that will give several colour changes over a fairly wide range of values. The 'Lovibond Comparator' is a cheap, simple piece of equipment, designed for the layman to determine pH values colorimetrically.

(b) Electrometric: A unit is available for determining the pH value of solutions, and shows a direct reading on a meter. This unit gives very accurate results, but is rather more expensive.

EFFECTS OF pH ON BISCUITS

The pH of a biscuit, both in dough form and when baked, affects several factors.

Flavour: Most flavours, particularly fruit flavours, are associated with, and appreciated better in, an acid medium. For the best flavour appreciation pH 6·0 is the optimum. It would be wrong, therefore, to produce a filling cream flavoured with lemon and having an alkaline pH. It is usual, of course, to add citric acid to a lemon filling cream so that the acid complements the lemon oil. In the case of chocolate and cocoa an alkaline pH 7·0-7·8 is considered best.

Fermentation: For optimum yeast activity a pH 5·0-5·5 is required. During fermentation of doughs this is seldom reached, and more often it will be between pH 5·5-6·0. If the pH is higher than 6·5, the speed of fermentation will be considerably retarded. If greater than pH 7·0, fermentation will probably not proceed at all.

Biscuit spread: Biscuits with a pH 6·0 will tend to hold their size during baking (although sugar particle size and quantity

AERATING AGENTS 49

also influence this factor) and as pH increases then the biscuit tends to spread correspondingly. This factor is considered to be controlled by the astringent effect of acid on the gluten structure of the biscuit; and by the softening effect of an alkali on the gluten structure.

Stability or keeping qualities: Biscuits with a pH 7·0 and above have better keeping qualities than those below pH 7·0. This is undoubtedly due to the fats being more readily attacked and broken down in an acid medium than in an alkaline medium. When the fats are broken down at all, rancidity quickly ensues. It is important, therefore, that very stable fats are used in biscuits of an acid nature.

Bacteria and mould control: Mould growth and certain bacterial actions can be inhibited by a fairly low pH, e.g. 5·0. The moisture content in biscuits is normally too low anyway, but in marshmallows and jam and jelly filling, deterioration or contamination by mould and bacteria can cause a great deal of trouble.

Dairy products

CHAPTER 5

In this chapter we shall deal with milk, cheese, and eggs. Butter, of course, has already been discussed in Chapter 2.

MILK

Although milk is approximately 87·0% water, it is a remarkably well balanced and nutritious food, consequently, it will enrich any product in which it is used in place of water. It will enhance flavour and because of the sugar present, it will increase the baking colour. Milk can be used to advantage in most biscuit doughs, and in biscuit fillings and coatings.

In bakeries, milk is rarely used in its fresh state owing to its bulk for storage, the variations in supply and in composition, and because of its poor keeping qualities. It is used in one of its preserved forms. Pasteurisation and sterilisation of milk help preserve the milk but still leave the disadvantage of bulk, so it is more often used in condensed or powder form.

As cream and butter are such valuable by-products of milk production, there is always a huge surplus of milk which has had the cream separated from it. This is known as skimmed milk. Milk products are available in full-cream form and in skimmed-milk form.

Condensed milk

Condensed milk is usually full cream. It is pre-heated to destroy bacteria and enzymes and is then evaporated to about ⅓ of its original volume in vacuum pans at a temperature of 65°C (150°F). This low temperature has little effect upon the flavour of the milk. About 40% sugar (sucrose) is added, which improves the keeping qualities.

Evaporated milk

Milk (unsweetened) is evaporated under vacuum in a similar

manner to condensed milk to about ⅓ of its original volume. The milk is then filled into tins and sterilised under pressure at temperatures up to 110°C (230°F). This high temperature treatment causes a considerable deterioration of flavour, but ensures a completely sterile product. Evaporated milk is available both as full cream and as skimmed. Unsterilised evaporated milk is available in milk churns, if the user is situated close to the processing plant. This product has only a limited life, but does not suffer the flavour deterioration of the sterilised version.

Dried milk powders

Milk powders are available either skimmed or full cream and can be processed by roller process or by spray drying.

(a) Roller process: A proportion of the water is removed by vacuum evaporation and the thickened milk is then run on to a heated drum which is revolving in a chamber either at atmospheric pressure or under vacuum. The remaining moisture evaporates, leaving the milk solids as a thin layer on the revolving roller. This layer is scraped off and ground to a powder.

(b) Spray process: Milk is partially condensed in vacuum pans and is then sprayed under great pressure through a fine atomising nozzle into a large heated chamber, through which passes a current of warm dry air. This completes the evaporation and the dry milk powder falls to the bottom of the chamber. This method gives a superior grade of powder and is more widely used.

Skimmed milk powder has better keeping properties than full-cream milk powder, as the fat content in the latter tends to turn rancid. The full-cream milk powder has a far superior flavour, but is more expensive.

Reconstitution of milk powders

1 lb full-cream milk powder should be added to 8 lb water and well whisked together. 1 lb skimmed-milk powder is added to

10 lb water (1 gal) and well whisked together to reconstitute. There is always a danger of the milk powder lumping if added direct to the biscuit dough, and there is also a danger of little dark-brown specks appearing on the biscuit face when added dry. Both these troubles can be avoided by reconstituting, or by blending the milk powder well with the fat content before any water comes in contact with the mixing.

Skimmed-milk powder can be used at the rate of 2-4% based on flour weight in biscuit doughs, and from 5-10% total weight in biscuit fillings. Full-cream milk powder can be used at half these rates.

Storage of milk powder

Milk powder should be kept cool and dry; it should be protected from the atmosphere, as the powder will absorb moisture and form lumps. (The lumps must be well ground down before using, as they will not break down in a dough or filling cream.) If possible, milk powder should be stored in air-tight containers.

Buttermilk

Buttermilk is a watery liquid which remains after the butter has been separated from the cream. During the ripening of the cream before churning, the lactose is almost completely converted to lactic acid and so buttermilk is only really a weak solution of lactic acid with traces of fat, protein, lactose, and mineral salts. Lactic acid has a softening effect on gluten, and if used in fermented or hard doughs, would assist in developing a mature gluten. As a high proportion of butter is manufactured from unripened cream, in which the lactose has not been fermented into lactic acid, most of the butter milk available is only very slightly acid or is neutral. It follows, therefore, that there can be little effect on the pH of the dough, but owing to the high lactose content, buttermilk can be used as a milk substitute. Buttermilk is available in powder form, which is better from a storage and handling point of view.

TABLE 3: *Typical analyses of milk and milk products*

	Fresh milk	Condensed Sweetened	Full cream Unsweetened	Evaporated Sweetened	Skimmed Unsweetened	Milk powder Full cream	Skimmed
Water	87·0	25·0	69·5	30·0	49·9	2·75	2·5
Fat	3·7	11·0	9·25	1·75	2·75	27·7	0·85
Protein { casein, albumen }	3·6	8·0	9·9	11·0	18·1	27·3	36·15
Sugar (lactose)	6·0	14·0	9·6	16·25	25·5	36·5	51·8
Mineral salts	0·7	2·4	1·75	2·25	3·75	5·75	8·7
Added sugar (sucrose)	—	39·6	—	38·75	—	—	—
Total	100·0	100·0	100·0	100·0	100·0	100·0	100·0

Whey powder

When milk turns sour, the lactic acid that is developed coagulates the protein matter (casein) to form curds. In cheese making, the curds are separated from the liquid phase which is known as whey. Whey is available in powder form and can be used as an inferior substitute for skimmed-milk powder. It is approximately 70·0% lactose, the remainder being moisture, protein, mineral salts, and a small proportion of fat.

CHEESE

Cheese is made from soured milk by separating the curds from the whey. The milk may be full-cream, skimmed or may even have added cream. The souring or curdling may be brought about by the natural development and activity of lactic acid bacteria breaking down the lactose into lactic acid, or by the addition of rennet. Rennet is an enzyme preparation from the digestive stomach of the calf. In hard cheese, the whey is separated from the curd, but in soft cheeses, a good proportion is left in. The curd is then salted and the cheeses made up and left to mature and for flavour to develop. The flavour is dependent upon the action of bacteria, either added as a culture, or finding their own way from the atmosphere to the interior of the cheese. The maturing period is carried out at carefully controlled temperatures. Blue- and green-veined cheeses have additional flavour developed owing to the presence of the mould causing the veins. Cheese made from full-cream milk consists of almost equal quantities of water, fat, and protein.

Among the cheeses suitable for baking purposes are: Derby, Leicester, Dunlop, Edam, Gouda; and English, Canadian, Dutch, New Zealand, and Australian Cheddar. For filling purposes, the more expensive and speciality type flavours are available, such as Stilton, Blue Danish, Gorgonzola, and Parmesan. There are also processed cheeses available, usually of the cheddar type, which when ripe, are ground up and pasteurised.

When using fresh cheeses it is important that they are used in the prime of condition. Storage in the factory is important and depends chiefly on the stage of the cheese when bought, and also on the type of cheese. Storage at 4·5°C (40°F) will keep ripe cheeses in a reasonable state for up to a month.

DAIRY PRODUCTS 55

For incorporation in a dough, the cheeses should have the rind removed and can then be broken down in a mixing machine to a suitable consistency. As cheese is used as a flavouring material, it is important to recall this when buying and not to buy on price alone. Consistency of flavour is equally important, so that a standard weight can be used.

This is one of the advantages of the processed cheese, as experts blend the cheeses prior to pasteurising and drying, and so a consistent flavour is maintained; the cheese is available in powder or as a fine meal, and no preparation is required before addition to the dough, and also, being pasteurised and dried, the keeping qualities are excellent. The disadvantages are standardisation of flavour, and the costs are higher.

Cheese is an acid ingredient, and for optimum flavour the finished biscuit or filling should be acid (pH 6·0). Pepper is complementary to cheese, as it accentuates the flavour. It should not be used to the extent of masking the flavour, or as a substitute.

EGGS AND EGG PRODUCTS

Eggs are highly nutritious, and therefore enrich any recipe in which they are included. Because of their albumen content they assist in structure formation, both in dough form and when baked. Eggs also appear to have beneficial effects on the crispness, the texture, and the eating qualities of biscuits. They also enhance the biscuit colour and flavour, and help to prevent wafers sticking to the plates and biscuits sticking to the oven band.

Quantities up to 5% based on flour weight can be used to advantage. Although the use of eggs is claimed to be advantageous, very few are now used in biscuit production, except in speciality products, owing to their relatively high cost. Soya flour is frequently used to replace eggs, or increased milk powder, but even if these products give similar handling qualities and eye appeal in baked goods, it is doubtful if the flavour can be recaptured.

Eggs are available in shell form, frozen, and dried as a powder. Unless treated, shell eggs will only keep a limited time, as the shell is porous and the membrane lining is semi-permeable. Bacteria are able to penetrate to the interior of the egg, and as the egg is broken down by the bacteria, strong smelling gases are given off. Shell egg treatment is designed to prevent the infiltration of bacteria and usually consists of forming an air-tight barrier around the egg. This can be done by storage in a 10% solution of sodium silicate (water glass), which will preserve for up to a year. Storage in lime water will preserve for up to 8 months. Coating the shells with a layer of varnish, lacquer, or wax, will seal the pores. Chilling will preserve eggs by retarding bacterial action. A temperature very close to, but slightly above freezing point, is necessary. If eggs are frozen in the shell, the water expands on freezing and cracks the shell.

Shell eggs should always be carefully inspected when cracking, as one bad egg can ruin a large batch if inadvertently mixed. It is sometimes economic to buy shell eggs, but cracking labour must be added to the cost for a fair comparison with other forms. Eggs that have been cracked in excess of requirements can be preserved for a day or two by the addition of sugar, at least 25% of the weight of eggs must be added and this addition must be accounted for when using the eggs in mixings.

Shell eggs should be kept in a cool dry store and used strictly in rotation. To test a shell egg for soundness, immerse it in a 10% brine solution. If the egg sinks, it is fresh, if it is bad or going bad, it will float, owing to the presence of the gases produced by the bacteria. If the egg is ageing, the moisture content of the egg will have decreased through evaporation and the air in the space between the shell and the membrane will increase, causing the egg neither to sink nor float. Candling is another method of testing shell eggs. The egg is held against an aperture in a frame, through which a strong light shines. If the egg has a clear and transparent appearance, it is fresh, if cloudy or opaque, then the egg should be rejected.

Frozen eggs are probably the most consistent source of eggs. Shell eggs are cracked and blended together, placed in tins and then quickly deep frozen. They are then placed in cold storage until required. Defrosting of frozen eggs is important, as once they are defrosted they will begin to deteriorate and will soon putrefy. Frozen eggs must not be removed from cold storage until a few days before use, then they should be stored in a refrigerator just above freezing point. if the egg is required in emergency, heat should not be applied. The quickest method of defrosting is to stand the opened tin in a stream of running cold water. Defrosted egg will keep for a day or two in a refrigerator quite satisfactorily, but if it has to be kept longer, then sugar should be added.

Dried egg powder is prepared by the spray-drying process, as for milk, but the temperature must not exceed $52 \cdot 5°C$ ($125°F$), otherwise the albumen content may coagulate. (The actual air temperature in the drying chamber may be as high as $176 \cdot 7°C$ ($350°F$), but owing to the high latent heat requirement of the evaporating water, the egg solids remain at a low temperature.) Dried egg will keep for considerable periods in a cool, dry store, preferably in air-tight containers. It is reconstituted by adding 1 lb sieved powder to 3 lb water and mixing well. For the best results, the reconstituted egg should stand for at least 30 min before use. The flavour is inferior to either frozen or shell egg.

Accelerated freeze dried (A.F.D.) egg is now available at prices competitive with frozen egg. It is produced using liquid egg from which the sugar content has been extracted by enzymic activity, and to which a stabiliser is added. After homogenising and cooling, the egg is placed in shallow trays in racks. The racks of egg are then blast frozen at approximately $-34 \cdot 4°C$ ($-30°F$). This is followed by extreme conditions of high vacuum. Gentle heat is then applied, causing the ice crystals to change from being solid directly into being a vapour. This process is known as sublimation. The dried egg is in the form of a finely honeycombed, or porous, cake which is finely broken and sieved to form a powder ready for packing. A.F.D. egg readily reconstitutes at the rate of 1 : 3 with water, in many cases, however, it can be used in the dry form. Normal storage conditions are suitable for A.F.D. egg, i.e. cool and dry; cold storage is not necessary.

Egg yolks are particularly rich and would enhance the flavour and appearance of most biscuits. The yolk is rich in lecithin, which is valuable for its emulsifying properties in doughs. Yolks are available both frozen and dried. The dried is reconstituted in equal quantities of powder and water.

Egg whites or albumen are valuable because of their ability to form a foam (or aerosol). They can be used in the manufacture of marshmallows and icings, but owing to the cost, gelatine is frequently substituted. Albumen is available frozen and in two dried forms. The frozen albumen should be handled in a similar manner to frozen egg. Dried albumen is prepared as crystal or powdered albumen. The crystal form requires approximately

8 hr soaking before use, whereas the powdered variety is ready for use within 30 min of reconstitution. Both forms are reconstituted at the rate of 1 part to 6½ parts water. When using albumen to produce a foam, it is important that all the equipment is completely greasefree, as grease, even in very small quantities, will prevent aeration.

Owing to the incidence of food-poisoning bacteria, such as salmonella, in eggs (particularly in imported eggs and egg products), legislation has been introduced in the United Kingdom making the pasteurisation of all bulk eggs compulsory. This permits shell eggs to remain unpasteurised, and when used on the premises still not to require pasteurisation where cracked out. All other forms of egg and egg products must undergo a pasteurising process before they are permitted for use in the preparation of food for human consumption. The pasteurising process consists of heating the egg to 64·4°C (148°F), at which temperature it is held for 2½ min and then rapidly cooled to a holding temperature of 3·3°C (38°F), prior to further processing. To prevent coagulation of the egg proteins, the temperature and time control must be precise, otherwise the quality will deteriorate rapidly. Even so, problems can arise, as the fat content of the yolk tends to break out of emulsion and causes difficulties, particularly when the egg is required for whisking. Because of this partial breakdown carefully defrosted pasteurised frozen egg gives the best results if thoroughly blended or homogenised before use.

Great care must be taken when handling egg and egg products, as even if previously pasteurised, contamination can easily occur, and while thorough baking will re-pasteurise the product, some products are not baked; and in other cases toxins will develop in the stored egg and these will not be destroyed by baking.

Egg substitutes are available in powder form, but these are chiefly prepared from soya flour, cornflour, and other starches, from acid and alkali chemicals for aeration, and colouring and milk powder or albumen. Although these substances may perform in a similar manner to the egg product they represent, they do not match up to the nutritive value or the flavour of the real thing.

CHAPTER 6
Fruits and nuts

THERE are numerous dried and preserved fruits and nuts and nut products available to the biscuit manufacturer. They can be used to advantage to produce varieties of flavours and finishes.

DRIED FRUITS

The majority of the dried fruits require a Mediterranean-type climate and are grown in California, Australia, South America, and South Africa, as well as those countries around the Mediterranean sea. They need rain in the winter and spring, followed by hot sun in summer and autumn. Raisins, currants, and sultanas are all products of different types of grape vine.

Currants

Currants are small black grapes that are either sun dried or shade dried by hanging on wires in sheds open at the sides to permit the free passage of warm dry air. Although this method is slower than drying in the sun, there is no danger of damage from rain, and as they are hanging up there is less contamination from stones and other undesirable items from the earth. The poorer quality fruit is dried by placing the bunches of grapes on mats in the sun and turning periodically. This method takes from 10-12 days, but if there is bad weather the crop can be ruined. When the currants are dry they are freed of stalks and stones and then packed. Greek currants are usually in 50-lb cases and sometimes in 28-lb or 35-lb cartons.
Good quality currants should be bold, fleshy, and clean, of even size and blue-black in colour. There should be no red shrivelled fruit present, as these have a strong acid taste and can spoil the flavour of the rest.
The new Greek crop of currants arrives in Britain in November. It is graded for size as bold, medium, and small. The medium and small are most suited to bakery uses, although the general quality of the small is lower than the medium. The grading for quality depends on type, and the best quality

currant is the Vostizza. Second quality currants, such as Gulf, should be quite adequate for biscuit manufacture. Zante are third quality currants. Patras and Amelias are low quality, but may be satisfactory according to the use for which they are intended. Pyrgos and Calamata are the poorest quality. They are less clean, have an abundance of stones, but are the cheapest to buy. It is possible that this last grade are sun dried on the earth without even mats, hence the inclusion of dust and stones.

Australian currants arrive in Britain in April or May, but only in small quantities. They are graded according to quality (1 to 5 crowns—5 crowns being best quality) and not according to size. All Australian fruit (including sultanas and raisins) is clean, as it is dried on racks and not on the ground, and a reputation for consistency of quality and reliability is being built up. Clean fruit means free of stalks and stones, but does not mean washed.

Sultanas

The bunches of seedless yellow grapes are dipped in a bath of potash and rosemary or lavender, which is covered with a layer of olive oil. This process softens the skins, brightens them and sterilises them. The bunches are then dried (either shade or sun according to quality). Sultanas labelled 'natural' are unbleached, and the majority of sultanas after being dried are exposed to the fumes of burning sulphur (SO_2), which bleaches the skins and also acts as a preservative. Sultanas should have a good flavour and colour, and be fleshy.

Greek and Cretan sultanas arrive in Britain in September to October packed in 28-lb or 35-lb boxes or cartons. They tend to become sugary by January. Grading is only by size from zero to 4, although the Cretan numbers vary. They are all good quality fruit, but are expensive. Turkish sultanas, usually known as Smyrna, arrive in Britain in September to October and are graded as Goldens and 11, 10, 9, and 8 in descending quality. Goldens are good quality fruit, but there is always considerable danger of stones, stalks, and other foreign bodies in all grades. Some Persian (Iranian) sultanas are of a very high standard, but in general, Afghan fruit should always be avoided, as there is more seed than fruit.

The Australian crop arrives in Britain between May and July, packed in 60-lb cases and 30-lb cartons. It is clean and chiefly bleached; like currants, sultanas are graded from 1 to 7 crowns in ascending quality, but are also graded according to size as 'large', 'small', and 'unsifted'; 7 crowns are very rare, 4 and 5 crowns are widely used for manufacturing. A small quantity of South African sultanas is imported. They are packed in 25-lb boxes and are graded from 1 to 5 diamonds in ascending order of quality. They are sometimes inclined to be rather acid to taste.

Raisins

The true raisin (or muscatel), is a large brown fruit with an excellent flavour, unfortunately, it contains rather large seeds, which have to be extracted. The bunches of grapes are partly dried by twisting the stalks while still on the vine and are then finished off in open sheds. When dry, the seeds are removed. Best quality raisins are produced in Valencia, Alicanti, and in Italy. Australian seeded raisins are of excellent quality, South African raisins are rather lean looking, but are becoming popular.

Californian seedless raisins (also known as Thompson's Naturals) are a small, very dark fruit, with a strong flavour. They have not the fine flavour of the muscatel, but are very tasty in their own way.

Dried fruits in biscuit doughs

It is usually necessary to wash dried fruits before use, and if a fruit washing machine is not available, then the fruit should be immersed in a liberal amount of water and swirled round for only about 1 min. The fruit is then put to drain in a sieve. The washing may be repeated again if the fruit is very dirty, but care must be taken that the fruit does not absorb too much water and become 'squashy', or the fruit will break down during mixing and discolour the dough. Too long soaking also spoils the flavour. After draining, the fruit should be carefully picked over, preferably by spreading the fruit on a tray and passing it by hand to a metal tray set at a level 3-4 in lower. Any stones or metal that are not seen should be heard as they fall on to the lower tray. If possible, the fruit should be spread out thinly on trays in a warm, dry place for 2-3 days, so that the excess moisture will dry off.

As biscuit doughs have to withstand fairly heavy machining and the biscuits are usually thin, fruit should be added to the mixing at the latest possible time to ensure even distribution throughout the dough with minimum damage; and the fruit should be the smallest available to suit the quality and price. One other advantage of using small fruit is the number of units per pound. Where a fruity flavour is required and the state of the fruit is immaterial in the finished biscuit, it may be advisable to mince the fruit before adding to the dough. In this case the fruit can be added at the creaming stage and certainly simplifies production and utilises the flavour of the fruit to the best advantage.

Figs

Figs are grown in the countries bordering on the Mediterranean and also in California. Smyrna figs are considered to be of the best quality and flavour. Calimyrna are a similar fig, grown in California. White Adriatics are a lower priced fig. Fig pastes of various blends are usually the most economical buy, as these are prepared from figs that are unsuitable for sale as whole fruit. Quality will vary considerably according to price, but most pastes are suitable for fillings in fig roll biscuits. There is nothing to be gained in using bleached fruit, as the colour darkens during the preparation of the filling and during baking. The figs will generally require washing before use, but soaking should be avoided, as sugar and flavour are lost. Storage of dried figs should be in a cool, dry atmosphere, at a constant temperature.

Dates

Dates are sun dried fruits of Iraqi and North African palms. They are very sweet and rich in sugar and could be used to enhance or vary biscuits containing minced fruit. For efficient distribution in a dough they should be soaked in about half their own weight of water for an hour or more until they soften.

FRUITS AND NUTS

SUGAR PRESERVED FRUITS AND PEELS

Candied citrus peels

Citrus fruits such as lemons, oranges, and citrons, grow in the Mediterranean countries, and other places of similar climate. Only the skin or peel of the fruits is used, and each fruit follows a similar process of treatment. Thick-rind fruits are cut across the middle and the pulp is removed. The halves, known as 'caps', are soaked in brine for several days to open the pores and to remove the acid taste. After draining and washing, the caps are placed in a tank of warm sugar solution. Syrup is absorbed by the caps, which pass progressively through stronger solutions of sugar until sufficient sugar is absorbed to preserve the skins. If the caps are to be sold 'cut' or 'sliced' or 'mixed', they are then allowed to drain and dry before being cut as required by machine. If candied caps are required, then they are drained and placed on wire trays in a warm chamber to dry quickly, which causes some of the sugar to crystallise. The candied caps generally retain their better flavour longer than the cut varieties. The peel should have a good colour and flavour, and should not lose its colour. Citron peel is very thick and consequently, a longer process of treatment is necessary. Peel should be stored in a cool dry place. Opened boxes should be well fastened down again or the peel will dry out and become tough and uneatable.

Cherries

Cherries are grown mainly in Europe, and the bulk of those used in Britain are exported from France. They are available as 'glacé' and as 'crystallised'.

(a) Glacé: Good quality fruit is bleached for several weeks in a solution of water, calcium carbonate, and sulphur dioxide, until the fruit is quite colourless. The fruit is stoned and washed. To soften the skins and flesh, the cherries are cooked for a few minutes and are then drained. The cherries are immersed in a weak syrup which is coloured, usually red, but also green or yellow. The syrup strength is increased daily by boiling over a period of about 9 days. The cherries remain in the syrup a further 5 or 6 days before draining and packing, with sufficient

clear syrup to preserve them. The packs weigh 11 lb (5 kg). Very small cherries are known as 'Alpine' cherries.

(b) Crystallised: After draining, preserved glacé cherries are rolled in fine castor sugar and dried in a warm chamber to crystallise in a similar manner to candied peel.

Cherries when whole, are normally too large for biscuit making, but if chopped up, their bright cheerful colour can be used to advantage.

Crystallised fruits

Pineapple, peaches, apricots, plums, pears, and other dessert-type fruits, either whole or in pieces, can be preserved in a similar manner to cherries and then crystallised. They have little application in commercial biscuit making, but could be used to create speciality lines in festive packs.

Angelica

Angelica is a large green plant grown in Belgium, France, and England, of which only the stem is used. It is preserved in a similar manner to cherries, using green syrup. Apart from the bright green colour it has an aromatic flavour.

Ginger-root

Ginger is associated chiefly with India and China, but it is also grown in Australia, Africa, and the West Indies. Only the tuberous root of the plant is used. It is well-washed and boiled in a weak sugar solution until soft. The syrup strength is gradually increased, as in the candying process of citrus peels. It is stored in syrup as root, chips, or crushed. It is also available crystallised.

Crystallised flowers

Violet and lilac flowers, and rose petals, are laid out on wires and a suitably coloured syrup is allowed to drip on to them.

When thoroughly saturated, the petals are dried over gentle heat. They can be used as a decoration, owing to the colours, but they also have a scented flavour of the original flower.

NUTS AND NUT PRODUCTS

Nuts are very useful ingredients for bringing about changes in flavour, texture, bite, and appearance in biscuits. There are several nuts available in various shapes or forms. They are all of a high food value, but unfortunately most of them are rather too expensive for the everyday run of biscuits.

Coconut

Coconut is the cheapest form of nut available, and fortunately, it is very popular with the biscuit-buying public. It is the white fleshy portion of the fruit of the coconut palm. The main sources of coconut are Ceylon, India, the West Indies, and the South Sea Islands. Coconuts are used widely as a source of oil for margarine and biscuit manufacture. The white flesh is known as 'copra'. This is removed from the shell of the nut and is dried either in the sun or in the shade. A better colour is produced by shade drying. When dried, the copra is cut according to requirements. The main varieties are the fine, medium, and coarse desiccated coconut, but it is also available shredded in two or three grades. A very fine grade of coconut is known as coconut flour.

It is particularly useful in the fine and medium grades for incorporation in biscuit doughs and in the finer grades for filling creams. The medium and coarse coconut can be used as a decoration. As a decoration, an improved flavour is obtained by lightly roasting to a golden brown colour. Coconut can also be coloured by well mixing with a liquid colour and then drying off the excess moisture.

Owing to the high oil content, coconut is liable to develop rancidity, but under good storage conditions of cool, dry airiness, with strict rotation of stocks, it should keep for quite lengthy periods. Coconut has frequently been found to be contaminated with salmonella bacteria, owing to the somewhat unhygienic treatment in the country of origin, and consequently, it can cause food poisoning. Under normal baking conditions it is unlikely that any bacteria will survive, but where raw coconut is used, it is essential that the pasteurised variety is used. Pasteurisation, of course, adds to the cost and tends to make the coconut a little oily, but these are minor disadvantages when compared with the dangers involved.

The outer fibrous husk of the coconut is used to make matting, and the shell is sometimes roasted and ground to a powder, which is then used to cheapen spices.

Almonds

There are two types of almond: the sweet and the bitter. The bitter almond is used almost exclusively for the production of essential oil of almond. The almond itself is not suitable for eating on its own, but may be used to boost flavour by blending a proportion with sweet almonds. Sweet almonds are grown mainly in Spain and the middle eastern countries of Asia. The kernel is enclosed in a tough brown skin, which is removed by immersing the nut in boiling water and then by friction between two canvas webs. After drying, the nut is ready for machine cutting into one of several shapes. The almonds are available whole (either blanched or unblanched), split (halves), flaked, strip (match shaped), nibbed (small dice), and ground. Almonds are also available in the form of marzipan and almond paste and macaroon pastes. A type of marzipan is also produced from the nuts of apricots and peaches. It is cheaper than almond marzipan and is usually called a kernel paste. In spite of their richness and fineness of flavour, because of their high cost almonds are hardly ever used in biscuit production, except in speciality lines, either in the biscuits or as decoration.

Walnuts

The walnut is a rather large wizened nut, grown chiefly in Southern Europe and in India. The skin is bitter, but can be removed by blanching as with almonds. Walnuts are available as halves and as broken. The flavour is quite strong. Walnuts have a high fat content which will turn rancid after prolonged storage.

Brazil nuts

Brazil nuts are grown in northern South America as well as Brazil. They are a quite large 'orange-segment' shaped nut. The flavour is distinctive and popular. The nuts are available whole, flaked, and sifted. They have a high fat content and are subject to rancidity during storage.

Groundnuts (monkey nuts, or peanuts)

The groundnut, a product of Mediterranean-type climates, is a small creamy white nut with an easily removed reddish-brown skin. Its flavour is not particularly strong, and the nut can, in fact, be treated to produce a neutral flavour. It is widely used in this way to adulterate, and to be a substitute for, more expensive nuts such as almonds. It is available whole, ground, and flaked. It has relatively good keeping qualities.

Pistachio nuts

Pistachio nuts are rarely seen in Britain, owing to their excessive price. They are sometimes known as 'green almonds', because of their green-coloured flesh. They are about 1 in in length, and have a purplish-brown skin which can be removed by blanching. Their flavour is similar to, but rather stronger than, sweet almonds.

Hazel nuts

Hazel nuts are a European nut, often known in Britain as 'hedgenuts'. They are similar to Barcelonas and to Filberts, the main differences being shape and size. They have a distinctive flavour which is enhanced by roasting. They are available whole and ground, their main disadvantage is that the brown skin is not removed by blanching. The distinctive flavour is used to advantage in praline and nut pastes; both of which can be used as flavouring ingredients in filling creams. Praline is produced by blending hazelnuts (or almonds) with melted sugar and a little lemon juice and then grinding them to a paste under granite rollers. Some commercial pralines contain chocolate in the blend.

Cashew nuts

Cashew nuts are bean-shaped nuts from India, Malaya, and the West Indies, that have a very bland, almost insipid flavour. They are used chiefly as a substitute for almonds and are available flaked and ground.

CHAPTER 7

Setting materials

FOR the manufacture of jams, jellies, marshmallow, and icing, ingredients are required that will either form a framework to hold fine bubbles of air and cause aeration and thickening of a syrup, or to cause a fruit syrup (jam or jelly) to set as a firm gel. In the case of forming a framework to hold air in a solution, the use of albumen has already been referred to in the chapters on aeration (mechanical) and dairy products (eggs). It is a rather expensive ingredient and gelatine is usually substituted.

GELATINE

Gelatine is a refined form of glue obtained from the skin and bones of animals. Pure gelatine is a transparent brittle substance without colour, smell, or taste. As quality decreases, colour, smell, and taste increase. It is available in sheet (leaf) form and as a crystalline powder, and will keep indefinitely if it remains dry. When moistened, or if it becomes damp, it will deteriorate rapidly.

Placed in water, it swells and will absorb about ten times its own weight. A hot water solution of 1·0% or more will set on cooling. The strength of the gel depends upon the quantity and quality of the gelatine used. It may be necessary to use as high as a 5·0% solution to achieve the desired strength of gel. Prolonged boiling, or a series of high temperatures over 60°C (140°F), destroys the gelling properties. Fresh supplies of gelatine should always be tested at a standard solution against a control sample.

Gelatine is prepared by steaming bones and skin at a temperature of approximately 60°C (140°F) for several hours, and the liquor obtained is run into shallow vats to cool. The grease and fat is removed and the liquor is filtered through animal charcoal and clarified with calcium sulphate. The clear liquid is concentrated in vacuum pans to a 5·0% solution. This solution is run out into thin sheets and dried in a purified atmosphere. (Gelatine solutions make ideal media for the growth of micro-organisms and bacteria.) Leaf gelatine is generally the highest grade. Powdered gelatine is prepared by grinding the sheets.

In icing, where gelatine is used only for aeration, approximately 2·0% of the total water content, or 0·5% of the entire mixing, should be ample. The aeration is relatively slight, and the drying out of the moisture holds the texture created by beating. When used in marshmallow, the gelatine performs a dual function. First it forms the network for aeration and then, as the mallow cools, the gelatine sets and holds the structure as a rigid foam. The usual quantity of gelatine used in marshmallow is in the region of 2·0-2·5%, but this will depend upon the quality of the gelatine in use.

The strength of a gel formed from a sample of gelatine is quoted in °Bloom. Bloom strength ranges from weak at 125° to strong at 275°. The higher the bloom strength figure, the less gelatine is necessary to produce a gel of a given standard, but the toughness of the gel and consequently, the mallow, increases. For marshmallow production, a bloom strength figure between 175° and 200° is usually considered most satisfactory. A higher bloom figure gelatine is more difficult to dissolve than a lower one, and it will start to set more readily during depositing. As bloom strength increases the price also increases.

AGAR-AGAR

Agar-agar is an extremely powerful jellying agent, derived from a variety of seaweed indigenous to southern Asiatic waters. It is available in powdered or fibrous form. It is insoluble in cold water, but will absorb large quantities of water, and when dissolved in boiling water produces a slightly cloudy gel with a less tough nature than one formed from gelatine. A 0·5% solution of agar-agar will set to a reasonbly firm gel. Boiling is necessary to dissolve agar-agar, but boiling does not detract from its setting powers.

When albumen is used for aeration in the manufacture of marshmallow, it is usual to use agar-agar as the setting agent of the foam. Like gelatine, agar-agar can be used for stiffening rather syrupy jam, but the main use of agar-agar is in the production of piping jellies. Piping jellies are brightly coloured, fruit-flavoured jellies of a soft nature, capable of being piped or

SETTING MATERIALS

deposited through a tube for decorative purposes. They are also used in marshmallows in the place of jam.

PECTIN

Pectin is the setting agent used in the production of jams. It is present naturally, to a greater or lesser extent, in most fruits. Apples, gooseberries, and citrus fruits are rich in pectin; strawberries, cherries, and rhubarb are poor. It is a water soluble colloid which, when in solution in the presence of sugar and acid, is capable of forming a gel. To obtain a firm setting it is necessary that the fruit contains not only pectin, but also acid. Fruits that are deficient in either or both can be supplemented by the addition of acid or pectin, or by a second fruit that is rich in either or both. Pectin is available commercially in two forms. Apple pectin, produced in the cider districts, is a thick brown viscous liquid which should be guaranteed acid free. Citrus pectin is rather more expensive and is marketed as a flavourless, odourless white powder. Either tartaric or citric acids are suitable as acid complements for pectin syrups.

ISINGLASS

Isinglass is an extremely pure form of gelatine obtained from the swimming bladder of various fish. It dissolves in hot water and sets as a jelly when cool. It is not widely used, owing to its high price.

SODIUM ALGINATE

Sodium alginate is derived from seaweed and is capable of forming quite firm gels of sugar solutions. It possesses no aerating powers, but can be used with albumen in marshmallow or for stiffening syrupy jams.

IRISH MOSS OR CARRAGHEEN

Irish moss is also derived from seaweed and is widely used as a setting and stabilising agent in emulsions.

GUMS

Although there is little use for edible gums in general biscuit manufacture, gum arabic (also known as gum acacia) is often used in a sugar solution as a glaze on baked almond rout biscuits. Although the syrup would form a glaze on its own, it would remain sticky. The inclusion of gum arabic ensures a glaze with a dry surface. A glaze composed of 2 parts gum arabic and 20 parts sugar, dissolved in 100 parts boiling water, will be quite satisfactory for this purpose.

British gum or dextrin is produced by the acid hydrolysis of gelatinised starch. It is soluble in water and forms an adhesive or paste.

CHAPTER 8
Chocolate and cocoa products

CHOCOLATE and cocoa are produced from cocoa beans which are the fruit or seed of the cacao tree. The tree grows in tropical areas, the main sources being Ghana, Indonesia, Mexico, and the West Indies. The flavour, price, and nature of the beans varies from area to area and, according to the type of product required, the chocolate manufacturer will prepare a blend of beans.

Cocoa bean treatment

The beans grow in oval pods from which they are removed and placed in heaps or pits, or in large containers in which heat develops and fermentation takes place. The pulp liquifies and runs away during this period and the purplish-white beans turn brown. They are then spread out to dry. The thoroughness of the drying affects the keeping quality considerably. The beans are packed and exported to the consumer countries for processing into cocoa and chocolate.

Cocoa beans are cleaned by sieving, aspiration, and magnetic separation, and are then roasted. The roasting is carried out in large revolving drums at temperatures between 100°C and 135°C (212°F and 275°F). This process is very critical, as upon it depends the building up of the chocolate flavour and colour, and the removal of excess acid and moisture. When cool, the beans pass through rollers to produce cocoa nibs. This action loosens the husks, which are then separated by aspiration and the germ is extracted.

Cocoa nibs from various sources are blended to give the desired flavour and nature to the product, and they pass through rollers to be ground to a thick viscous liquid of a reddish-brown colour which, on cooling, sets to form 'cocoa mass'.

Cocoa production

The cocoa mass contains approximately 54% cocoa butter, and in the production of cocoa much of this has to be removed by mechanical expression. Under pressures up to 6 000 lb psi the cocoa butter content of cocoa is reduced to between 20 and 30%, depending on the grade. The expressed cocoa butter is available for various purposes. The cocoa press cake is then broken up, ground to a fine powder and sieved into grades. It may also be sweetened.

CHOCOLATE COUVERTURE PRODUCTION

Both sweetened and unsweetened couverture follow similar refining processes, the sweetened having pulverised sugar added up to 40% of the total.

The cocoa mass (and pulverised sugar) is ground and mixed in a 'melangeur', a machine which consists of two heavy revolving granite rollers in a rotating granite bed. The plasticised mixture then passes to a series of refining rollers which ensure that no gritty pieces of any kind remain, before proceeding to the conches. The conches are machines consisting of troughs in which heavy rollers on arms pass backwards and forwards. This process is to ensure a smooth, perfect blend and may continue for 24-72 hr.

The chocolate is then poured into moulds of the required shape and size and allowed to set ready for packing.

Unsweetened chocolate is very bitter, but is by far the best flavouring medium for filling creams, as it gives the true chocolate flavour.

Milk chocolate couverture has full cream milk solids added at the melangeur stage.

BAKERS' CHOCOLATE COMPOUND

Bakers' chocolate compound is produced from cocoa powder from which the maximum amount of cocoa butter has been extracted. The cocoa butter is then replaced by a hardened vegetable oil. By using a hardened vegetable oil, a chocolate can be produced with any specific melting point, and with an eating quality ranging from soft to a brittle snap. One other advantage is that 'tempering' of this type of chocolate is unnecessary. It is quite sufficient to ensure that the chocolate is completely melted (temperature of 43-49°C (110-120°F)) and it is ready to use. It will set up readily with a reasonable gloss.

CHOCOLATE AND COCOA PRODUCTS 75

The disadvantages are that, being made from cocoa, it tastes more like cocoa than chocolate; the finish is inclined to be a little greasy in appearance; the gloss is not as good as properly handled couverture and its shelf life is only 2 to 3 weeks before it discolours.

CHOCOLATE COATING

To produce a chocolate with a long shelf life, a good gloss, a better flavour than bakers' compound, and at a lower price than couverture, special stearine butters have been produced which are similar to cocoa butter. They are the higher melting fraction of hydrogenated vegetable oils, separated by fractional crystallisation from suitable solvents. The coating is produced by substituting the cocoa butter or a proportion of it by the stearine butter. This type of product can be used without tempering as for bakers' compound, but to achieve similar results to couverture it must be handled as couverture is handled.

Tempering of couverture

If couverture is melted and then allowed to set, it will have a cloudy or streaky surface. This effect is due to the different melting and setting points of the cocoa butter glycerides. Each glyceride crystallises when the chocolate reaches a certain temperature and crystals form slowly and at different temperatures causing 'fat bloom'. To avoid this and produce a surface that is clear and has a high gloss, it is necessary to accelerate the crystallisation of the glycerides. This is achieved by 'seeding', that is, the introduction of crystals which form a nucleus for subsequent crystallisation.

Tempering is the process by which this acceleration can be brought about. Couverture is melted at a temperature of 44°C (110°F), stirring continuously to ensure a completely homogeneous mixture. (Some manufacturers add lecithin during the production of chocolate and this will assist in obtaining a completely blended product.) The couverture is then cooled to 28°C (82°F), during which the higher melting fractions of the cocoa butter start to crystallise. If the couverture was kept at this temperature it would gradually set, but now that the 'seed' crystals are formed the temperature is raised to a working temperature of 30°C (86-87°F). The rise in temperature is insufficient to melt the seed crystals so that, after use, when the temperature falls again, rapid crystallisation occurs—presenting a clear bloom and a high gloss.

Owing to different production processes and blends the manufacturers may advise slightly different temperatures for tempering, but basically these previous temperatures are suitable for milk couverture. To temper plain couverture, temperatures 1°C (2°F) higher should be used.

Although it is not necessary to temper chocolate coating, it should be tempered to obtain the gloss and long shelf life of a couverture. The temperatures used are higher than for couverture and, depending upon the proportion and type of stearine butter in use, may vary from manufacturer to manufacturer. However, melting should be accomplished at 44°C (110°F), then the coating is cooled to 32°C (90°F), and finally raised to a working temperature of 37°C (99°F).

There are three basic methods of tempering couverture under large scale production conditions. The first is to melt the chocolate in one kettle and then transfer it to a second kettle to cool and reheat, from where it is fed directly and continuously to the enrober. The second method is to temper only the first batch of chocolate in the above manner, and thereafter constantly feed chocolate in at the same rate as it is being used. The incoming chocolate should have been melted as in the first, but only cooled to a temperature 1°C (2°F) above the temperature of that in the final kettle. The original seeds constantly form fresh seeds, and as long as the temperature and levels are scrupulously maintained, then the temper of the couverture will be maintained. The third method is a unit which automatically and continuously tempers a stream of chocolate as it is fed to the unit. The unit cools it in two stages and then reheats in the third stage.

If temperature control of the chocolate is so important to obtain first class products, it is just as vital to observe strict temperature control of the goods to be enrobed and of the enrobing room. For that matter there should be strict temperature control all along the line until the product is in the retailer's hands. Even he should try to maintain some control in his own, and the producer's interest.

The enrobing room temperature should be held at a constant 21°C (70°F). If possible, the biscuit bases should have been stored at a similar temperature for two days. If the bases are cold, they will chill the tempered chocolate and cause fat bloom. If they are too hot they will melt the chocolate, destroying the temper and again causing fat bloom. In addition to temperature control when enrobing, it is important to see that there is a minimum of biscuit dust and bits present, as these will enter the chocolate and cause thickening and lumpiness. Marshmallows should present a dry surface for the chocolate, otherwise the moisture will cause 'sugar bloom'. Some of the sugar in the chocolate dissolves in the moisture and later crystallises causing a whitish surface and sometimes pale fawn spots. With chocolate-covered marshmallows, it is important that the moisture content of the base is correctly controlled, otherwise splitting of the chocolate covering will occur.

Cooling of the chocolate should be achieved gradually so that when cooling tunnels are in use, the temperature at the feed end should be fairly close to room temperature and should fall gradually to its lowest point about three-quarters of the way along the total length. The final length should have a slight rise in temperature so that the chocolate temperature and packing room temperature differential is not too great. The actual temperature should be in the region of 18°C (65°F) at the feed end, falling to 14°C (58°F) at the coolest point and rising again to approximately 16°C (62°F). (Lower temperatures are possible depending upon ambient conditions.) In a tunnel where forced cool air is used for cooling, it should be introduced at the three-quarters stage (coolest point) and should move towards the feed end (against the current of biscuit direction). This will automatically achieve the fall in temperature towards the coolest point and then a slight rise over the final quarter. The minimum time for cooling is in the 12-15 minute range, the longer the cooling time, the better the finish will be; but in straight through tunnels, too much space would be required, with tiered (particularly overhead) types of coolers, periods up to an hour can be achieved. Too rapid cooling of products will cause discolouration through fat bloom. To prevent feet forming at the base of fully covered products, a chilled table should be used under the conveyor immediately after the enrober.

Packing room conditions should be draught free and maintained at a temperature 19°C (66°F). The relative humidity should be approximately 50%. The greatest danger in packing rooms is when the dew point rises above the temperature of the chocolate goods as they emerge from the cooling tunnel. If this happens, then condensation will form on the chocolate surface, resulting in sugar bloom.

Storage of chocolate should be at a constant temperature, preferably in the region of 18°C (65°F), if the temperature is frequently altering, the chocolate is liable to be unstable and it will lose its gloss and possibly develop fat bloom. If stock room temperatures are low and the stock is then transferred to a warm atmosphere, there is a real danger of condensation occurring, resulting in sugar bloom. The relative humidity should never exceed 70%. Retailers should be made aware of these dangers so that they understand the difficulties and can then make provision to counter them.

Chocolate bloom

There are two types of bloom to which chocolate is subject, one is fat bloom, the other is sugar bloom. Fat bloom is caused through the incorrect crystallisation of the cocoa butter (or even stearine butter) and may be the result of poor tempering; incorrect temperatures of: enrobing room, base to be coated, cooling tunnels, packing room, stock room, warehouse, or retailer's shop; or of packing before the chocolate is properly set.

Sugar bloom is generally due to the presence or contact of moisture. The sugar dissolves and then crystallises on the surface as a white bloom or even pale fawn spots. Contamination with moisture can occur by condensation on the chocolate during delivery or storage; from water or steam leaks in the melting or tempering kettles; or from water or steam leaks in the enrober. In the case of marshmallows, it may be due to the moist surface of the mallow. Condensation on the finished product may occur if the temperature of the chocolate is lower than the dew point. This applies to goods emerging from the cooling tunnel, or being transferred from a cold stock room to a warmer atmosphere.

CHOCOLATE AND COCOA PRODUCTS

Forms in which chocolate is available

The most common pack for chocolate is in block form, usually in 7-lb, 10-lb, or 14-lb moulds, these may be wrapped singly or be in cartons of 28 lb, 50 lb, or 56 lb. Chocolate is also available in bulk as a liquid delivered by tanker, and some firms are now supplying chocolate in crumb or chocolate drop form in 50-lb or 56-lb paper sacks (this saves breaking the chocolate up and reduces the time taken to melt, but it is bulkier and requires more storage space).

Uses of chocolate and cocoa

The main uses of chocolate are obviously as a finishing medium. Whether the products are half-coated or fully coated, they completely alter the appearance and taste of biscuits and so introduce new lines. Chocolate lines are also very popular with the general public, who appear to have been educated to prefer milk chocolate and consequently, plain chocolate is rather neglected.

Unsweetened or bitter chocolate is the ideal flavour in creams when a true chocolate flavour is required.

Cocoa powder can be used for flavouring creams and biscuit doughs, but it must be recalled that the flavour will be of cocoa and not chocolate. However, in baked products, this is not very noticeable. The flavour shows to best advantage in a neutral or slightly alkaline base. Between 5·0 and 10·0% cocoa powder, based on flour weight, is quite sufficient to produce a chocolate coloured and flavoured base. An uncoated biscuit that is to be named as a chocolate biscuit in the United Kingdom should contain a minimum of 3·0% fat-free dry cocoa solids in the finished product. To estimate this it is necessary to know:

(1) the baked weight of the mix,
(2) the moisture content of the cocoa powder, and
(3) the fat content of the cocoa powder.

The following calculation can then be used:

$$\frac{\text{baked weight of mix in lb} \times 3 \cdot 0}{\text{fat-free dry cocoa solids as a percentage}} = \text{Cocoa powder weight in lb at } \textit{minimum} \text{ legal limit}$$

To be on the safe side, this weight should be increased by multiplying by 1⅛ (1·125) at least, to the nearest amount suitable for ease of weighing.

Products not satisfying this requirement must not be called 'chocolate'; those having only a coating of chocolate must be called 'chocolate coated'.

CHAPTER 9

Flavouring materials

THE build up of flavour in biscuits is a very complex matter, depending on many ingredients which give 'background' flavour in addition to those used to give the definite dominant flavour. Flours from different sources and wheat strains have varying flavours, and as flour is the main ingredient, this must influence the final biscuit flavour. The use of dark sugars in place of refined sugars; of butter or margarine in place of a neutral blend of vegetable oils; the inclusion of fruit or nuts, of cocoa, of egg or cheese or milk, have already been discussed and all play their respective parts in producing flavour. It is the usual practice, however, to include salt to draw out flavour, and also to complement, boost, or change the natural flavour by the addition of some strongly flavoured substance such as a spice or an essence.

SALT

Common salt (sodium chloride) is a mineral product which is essential in small quantities for health and vitality. Happily, it is just as essential to our palate when consuming most foodstuffs. Biscuits made without salt do not lack flavour, but are insipid. (Biscuits with excess salt are, of course, also unpleasant to eat.) Although flavour is present in the biscuit, without salt it remains in the background and remains unnoticed. The addition of salt draws out the flavour, amplifies and enhances it. The quantity of salt required in a mixing varies according to the type of mix and the presence of strongly flavoured ingredients, but the usual quantity is in the region of 1·0% of the flour weight. Account must be taken of the presence of salt in other ingredients. If, for instance, margarine or butter is included in the mixing, then a reduction must be made in the salt content corresponding to the salt content of the fat used. (Butter and margarine have a salt content of approximately 2·0%.)

Apart from the influence salt has on flavour, it can also be used as a control of fermentation. If there is no salt present in a fermenting dough, fermentation will proceed rapidly. If the salt content exceeds 2·0% of flour quantity, then the fermentation rate will be retarded excessively, even to a point of stopping altogether if the salt content is too high. Generally, it will be found that if fermentation is brought to a halt by the salt content, apart from lacking the benefits of fermentation, the resulting biscuits would be too salty to eat. Salt also has an astringent effect on the gluten.

AUTOLYSED YEAST AND YEAST EXTRACTS

Yeast is autolysed, usually with salt, but sometimes by heat, or by hydrochloric acid which is later neutralised. Vegetable extract flavours and caramel colour may be added. The liquor is centrifuged and the extract may be concentrated under vacuum, or by roller drying. During autolysis the salt concentration is high to prevent the growth of putrefactive bacteria, and consequently yeast extracts are salty in taste. Dried autolysed yeasts are sometimes mixed with starch before drying.

Yeast extracts are used in artificial meat extracts and various savoury flavours, and can be used in most savoury biscuits and fillings. They should be used with care as the flavour can be very strong.

SPICES

Spices are aromatic vegetable products, usually used in powder form, to add strong characteristic flavours to food products. Owing to their strong flavour, spices are easily adulterated, when in the ground form, by the addition of the ground up shells of nuts, or of rice starch. A more difficult form of adulteration to detect is when spices have had the essential oils extracted to produce essences and the residue is ground up to dilute the spice. When buying spices it is good policy to buy from a reputable manufacturer and to make sample test bakes to check for flavour strength.

Ginger: Ginger is the most widely used spice in biscuit making and is very popular. It is produced from the tuberous root of a herb which grows in India, China, Nigeria, and Jamaica. The Jamaican ginger is generally considered to be the best, but the African is usually stronger and cheaper, and quite suitable for the biscuit industry.

The root is harvested when a year old and can be used to produce black or white ginger. Black ginger is the stronger in flavour, and is prepared by cleaning the roots which are then boiled, dried, and finally crushed and ground to a powder. White ginger is prepared from the core of the roots only. Instead of paring off the outer covering of the tuber, frequently the whole root is bleached and then treated in a similar manner to black ginger.

The starch of ginger is elliptical in shape and is transparent. It is easy to detect other starch adulterants by microscopic examination.

Other forms of ginger are available in sugar preserved forms: crystallised; root in syrup; crushed, and as ginger chips.

Cinnamon: Cinnamon is the bark of a species of laurel bush grown in Ceylon and Indonesia. The finest quality is from Ceylon. The bark-bearing shoots are cut when two years old and the bark is removed and dried. It is available in 'quill' form and is also ground to a powder. Cinnamon may also be distilled to yield its essential oil. The leaves are used to produce eugenol which, in turn, is used in the production of artifical vanilla essence.

Cassia: Cinnamon has always been an expensive spice whereas cassia closely resembles cinnamon in flavour and is much cheaper. It is produced from the bark of a similar bush, but the bark is rather coarser in appearance, aroma, and flavour. The best quality cassia is known as Saigon. Other qualities come from China, India, and Malaysia. Although it is used as a substitute for cinnamon, it is widely used in the preparation of mixed spices.

Cloves: Cloves are the flower buds of an evergreen shrub of the myrtle family which grows in Tanzania, Malagasy, and Indonesia. The buds are harvested just prior to opening and are sun dried. Cloves are not widely used in the baking industry, but yield up to 20·0% of their bulk as an essential oil, the main constituent of which is eugenol. They are, therefore, the chief source from which artificial vanilla essence is manufactured.

Nutmeg and mace: Nutmeg is the kernel or seed of the nutmeg tree which grows in Indonesia, Grenada, and Trinidad; mace is the fibrous network surrounding the shell containing the nutmeg. The mace is covered by a fleshy pericarp, the whole fruit being similar in build up to an apricot or peach. When the fruit is ripe, it splits to reveal the red mace which is removed from the kernel and, after being flattened, is sun dried. During the drying process the colour changes to a yellowish brown. Mace can be used as a spice in its own right, but it is more frequently mixed with ground nutmeg or in mixed spices.

The nutmeg is dried over slow fires or in the sun, and when dry enough, the shells are carefully removed. Nutmegs are available both whole and ground. Mace and nutmeg should be stored in air-tight containers as the essential oil is very volatile. Freshly ground nutmeg has a much richer flavour than the ready ground variety and is, of course, free from adulterants.

Allspice or pimento: The allspice, or pimento, is a berry from a tree closely related to the clove, and is grown in the West Indies. The reddish-brown berry, similar to a blackcurrant in size, is picked before it is fully ripe. If the berry is allowed to ripen, the pungency of the flavour is completely lost. The flavour of the berry resembles a mixture of cinnamon, nutmeg, and clove, which explains the name 'allspice' and consequently, it is valuable in mixed spices. The essential oil is rich in eugenol.

Pepper: Peppercorns are the seeds of small creeping vines which grow in Indonesia and the West Indies. They are available as black and as white pepper, either in uncrushed or in ground forms. Black pepper is from the under-ripe seeds and has the greater pungency and aroma. It is therefore the most satisfactory from the flavour point of view, but may not be so popular because of its appearance. In biscuit manufacture the appearance will generally be of minor importance. White pepper is the ripened seed which usually has two or three layers of husk removed. This produces a ground pepper of better appearance (particularly for table work), but of considerably diminished flavour. Freshly ground peppercorns are more pungent than the ready ground form.

Pepper is a useful flavouring material in the production of savoury fillings and biscuits, particularly in conjunction with cheese, as it accentuates the flavour. Other types of pepper are Cayenne and Paprika.

FLAVOURING MATERIALS

Cayenne pepper is prepared by grinding capsicum and chillies into a powder. Capsicums and chillies are the fruit pods of plants in the tomato family. When grown in the tropics, the fruits are very pungent, but when grown in Europe (Mediterranean climate) they are much milder.

Paprika is produced from a sweet red pepper or capsicum grown in Hungary and in the countries bordering the Mediterranean. Good quality paprika has little pungency, but is valuable for its strong reddish colour and its flavour.

Mustard: Mustard is the flour produced by grinding the seed of the mustard plant and may be either black or white. Best quality white mustard is grown in East Anglia. The black mustard seed is far stronger than the white in pungency and aroma. Mustard flour is produced from a blend of the two varieties.

Curry powder: Curry powder is a flavouring material composed of a variation of spices according to the manufacturer's own requirements. The spices normally used may include cardamom, chillies, coriander, cumin, fenugreek, ginger, nutmeg, pepper, and pimento. Turmeric is used for colour.

Caraway seeds: Caraway seeds are the aromatic seeds of a plant grown in Holland and other European countries. The seeds are about ¼ in long, curved and of a dark colour. They are used whole, ground up, or replaced by an alcoholic extract. They are also used in some mixed spices.

Coriander: Coriander is the yellowish-brown, hollow, globular seed of a plant grown in Italy and Mediterranean climates. It is used in mixed spices; to complement honey and ginger; and in curry powder.

Cardamom: Cardamoms are the small dark seeds of a plant of the ginger family which grow in India, Ceylon, Indonesia, and the West Indies.

Celery seed: Celery seed is the strongly aromatic seed of the celery plant which can be used in savoury products.

Poppy or maw seeds: Poppy seeds are small bluish-grey seeds, mainly of Polish origin, used on Continental and Jewish breads as a decoration. When crushed they are used as a filling.

Mixed spice: Mixed spices vary from manufacturer to manufacturer, but generally cinnamon is used as the dominant flavour. The other spices used may include cassia, allspice, ginger, nutmeg, mace, coriander, caraway, cloves, and pepper. The spices may be diluted by the addition of rice starch or by the use of spices from which the essential oils have been extracted.

Spice extracts: It is now possible to purchase the essential oils, or extracts of spices in liquid or paste form. These are extremely powerful flavouring materials, and care is necessary when handling them.

ESSENCES

Essences are widely used in the biscuit industry for flavouring all types of products. They are used to enhance the natural flavours and to give a definite flavour to the product. The term 'essence' is a loose descriptive word used for any powerfully flavoured substance (usually a liquid) which will impart its own flavour to another substance that is greater in volume. Essence can be divided into two groups: natural and synthetic.

Natural essences

Natural essences or flavours are obtained from the natural product which gives the essence its name. The group can be further sub-divided into:
   (1) natural extracts
   (2) essentials oils, and
   (3) true fruit essences or concentrates.

The natural extracts are prepared by mascerating the materials in ethyl alcohol, *iso*propyl alcohol, glycerine, or a mixture of these solvents.

Essential oils are chiefly responsible for the flavour and aroma of foodstuffs, and in some instances it is possible to extract the essential oil by expression or by steam distillation. Unlike the fixed or fatty oils, essential oils are completely volatile, and although they will make a greasy stain initially on paper, they eventually evaporate. Sometimes essential oils are dissolved in the previously mentioned solvents to dilute them or make them more readily soluble. In this case they would be classified as natural extracts.

The true fruit essences or concentrates are prepared by infusing fruit in alcohol and then concentrating or distilling to form a delicate flavoured fruit essence that is expensive and unsuitable to withstand high temperatures. They should only be used in high quality fillings where the flavour is shown to its best advantage and where it will not be damaged by heat.

Synthetic essences

Synthetic essences are those which are produced chemically from a source having no connection with the flavour it represents. They can be produced in two ways. It is possible to prepare the main flavouring component of an essence chemically. For instance, vanillin is the active flavouring substance in vanilla and can be produced by oxidation of eugenol from cloves. The two vanillins are chemically the same, but the synthetic product does not have the bouquet or subtle qualities of the natural vanillin. It is, however, much cheaper. The second way of synthesising an essence is by the reaction of an alcohol with an organic or inorganic acid. This product is known as an 'ester'. Esters are the base of many synthetic essences, particularly fruit essences. Many of the esters resemble certain fruits, but others can be imitated by blending two or more esters and possibly even some essential oils. If ethyl alcohol is reacted with butyric acid, the result is ethyl butyrate, which is very similar to pineapple in flavour.

Synthetic essences are very powerful and should be used with caution.

Although the classifications already given are generally accepted, it should be realised that essences may be prepared by combining any of the groups. A natural extract may be cheapened and strengthened by the addition of synthetic products. It must be remembered that most essences are volatile, so they should be kept in well-stoppered containers. Many are sensitive to light and should be kept in dark glass bottles. Others are very corrosive (lemon and orange), so should not be kept in tins, except for short periods, and even then the containers should be well tinned inside.

Vanilla

The natural extract of vanilla is prepared from the seed pod of the vanilla plant, which is a species of orchid growing in Mexico, the West Indies, and South America. The best quality vanilla pods come from Mexico. The pods are harvested slightly under-ripe and are 'cured' in the sun for approximately a month. During this period the active flavouring substance vanillin develops. The needle-shaped vanillin crystals can be seen on the exterior of the pods, and are also present inside the pods. Although vanillin is responsible for vanilla flavour, there are resins and gums which help to build up the fine bouquet and flavour of best quality vanilla extract. The extract is prepared by soaking the ground-up pods in alcohol. When all the flavouring components are completely dissolved, the alcohol extract is filtered off. It is usually coloured a little with caramel before bottling. Other grades may be diluted by the addition of sugar or glycerine, or may have synthetic vanillin added.

Synthetic vanillin can be produced more cheaply than the high quality extracts from vanilla pods, and owing to its similarity in flavour, it is used almost to the exclusion of the natural product in baked products. The extra fineness and subtlety of flavour of the natural product can be advantageous in creams, but it is very difficult to detect any difference in a biscuit after baking. The synthetic vanillin is prepared by the oxidation of eugenol. Eugenol is present in considerable proportions in clove oil, it is also present in cinnamon leaves, and in allspice or pimento. Vanillin can be purchased in the crystalline form or dissolved in alcohol and coloured as an essence.

Ethyl vanillin is a synthetic product with approximately three times the flavouring power of vanillin.

Almond

Sweet almonds contain only a fixed oil and no essential oil, whereas bitter almonds contain both. To obtain the essential oil of almond, bitter almonds, and also peach and apricot kernels, have the fixed oil removed by grinding and by expression. The press cake is allowed to stand, dispersed in water. During this period enzymic activity breaks down the amygdalin content

FLAVOURING MATERIALS

into benzaldehyde, glucose, and prussic acid. The poisonous prussic acid is rendered insoluble by the action of ferrous sulphate and milk of lime, and filtered off. The filtrate is distilled to yield the essential oil of bitter almonds—benzaldehyde. Benzaldehyde readily oxidises into the flavourless benzoic acid if in contact with the atmosphere. The essential oil is frequently dissolved in a suitable solvent to form almond essence.

Benzaldehyde can be prepared synthetically from a coal tar product 'toluene'. The synthetic product is chemically the same as the natural benzaldehyde, but has a less attractive bouquet and is harsher in flavour, owing to the presence of traces of other substances derived from the almonds in the natural product.

Lemon

The essential oil of lemon occurs in the glands of the rind and it is extracted by expression. The finest qualities are expressed by Mediterranean countries, and to some extent California. The essential oil consists of 4·0-5·0% limonene and citral, the remaining proportion being chiefly terpenes. Terpeneless oils can be obtained by fractional distillation and are frequently more than twenty times more powerful than the original oil, but are more expensive and do not dissolve too readily.

Essences are prepared by dissolving the essential oil in alcohol. Citral can be prepared from lemon grass oil and can be used to adulterate the essential oil or to prepare synthetic essences. The natural essence is superior in flavour and bouquet.

Orange

Apart from the fact that there are two essential oils of orange, bitter and sweet, the notes on lemon flavours apply in all details for orange flavours as well.

Orange and lemon flavours are also obtainable in paste form. These are excellent for flavour, but fairly large quantities are necessary and tend to be more expensive than the essential oils or other essences. They are prepared from finely ground peel and some may also contain juice, as well as added essential oil.

Peppermint

The essential oil of peppermint is obtained by distillation of the peppermint herb. English peppermint is the finest quality and is more expensive than the USA and oriental products. The essential oil is extremely pungent and is usually diluted with alcohol in the form of an essence. Peppermint should be treated with a great deal of respect on account of its strength, and also because its flavour will be readily picked up by other creams and biscuits in close proximity, or following on the same machine.

Spices

Essential oils and extracts of many spices and aromatic herbs and seeds are available. They are very potent and should be used with care.

Fruit flavours

Soft fruit flavours are not easily or satisfactorily prepared except as expensive extracts and concentrates. They are, however, readily reproduced by the use of esters, either singly or in combination. These flavours are generally rather harsh and not always accurate.

Other flavours can be prepared in the factory by dissolving the synthetic active flavouring material in a suitable solvent.

Butter flavour

So called 'butter' flavours are widely used, but are difficult to produce with any accuracy. Their inclusion in biscuit doughs is no doubt beneficial in the baked product, but does not compare with the flavour produced by the use of butter. Most of the flavours are based on the powerful flavouring agent diacetyl, which is generally diluted considerably in a suitable solvent, and which, even then, may require further dilution in the factory to ensure ease of handling.

FLAVOURING MATERIALS

Coffee extract

Coffee extracts are prepared by concentrating an infusion of water and freshly roasted ground coffee beans. Many extracts are available, some in liquid form and others as a powder or 'instant' coffee. They are frequently expensive and the cheaper ones will be diluted by the use of chicory to replace some of the coffee. When used in mixings they confer not only flavour but also colour.

Chocolate

The use of chocolate as a flavouring material has already been referred to but, wherever medium and price permit, unsweetened chocolate should be used. The only alternative is the use of cocoa, and when using cocoa, it should be recalled that the higher the cocoa butter content, the better the flavour will be. Chocolate flavours and compounds are available, but as these are based on cocoa and unsweetened chocolate, there seems little point in paying the extra costs entailed.

Liqueurs

Liqueurs are produced by soaking aromatic herbs, seeds, and spices in brandy until the flavours are extracted. The alcohol is distilled, sweetened, and matured in casks before bottling. They are, therefore, similar to essences (essential oils dissolved in alcohol) and can be used for flavouring purposes. Unfortunately, they are liable to heavy taxes and duties, and this prohibits their use except in the highest class speciality products. There are cheaper imitations of the original liqueurs, but the flavours of these are harsher. It is possible also, to obtain essences purporting to resemble various liqueurs and wines.

REQUIREMENTS OF ESSENCES

For ease of handling in the factory, essences should be of such a concentration that they can be readily measured and checked.

To achieve this it may be necessary to have the essences diluted or prepared in the laboratory by an operative capable of handling small quantities, or expensive commodities. When essences are to be used in doughs they should be readily soluble in water and should be stable during baking, that is, they should not break down or alter in any way by the action of heat, acid, or alkali.

Essences for use in creams should be readily dispersed in a fat and sugar mixture where there is no water present. Essences should not deteriorate during storage, nor should they be corrosive. As these factors are difficult to achieve it is usual to store essences in dark-coloured glass containers which are kept well stoppered.

Reference has already been made to flavouring materials in liquid and paste form. Many manufacturers are now producing them in powder form.

Uses of essences

Expensive, delicate flavours, should not be used in goods where there is any possibility of breakdown during baking or when in contact with the chemicals in the dough. They should be used where their excellent flavour will be recognised and appreciated to the full, and where the extra cost is merited by the profit margin. This type of flavour will be beneficial in marshmallows and filling creams where the use of cheap harsh flavours can spoil the sales. The use of cheap flavours is not necessarily advocated for baked products, as there is a good case for the use of medium priced essences. The object in view is to obtain the best possible flavour in the biscuit at a pre-determined cost. Cheap materials will give distorted flavours, and often a greater value is necessary to achieve the same degree of final flavour.

WHAT IS FLAVOUR?

Flavour is a sensation depending for its observation upon the senses of taste, touch, smell, and sight. Each of these senses analyses the product to be eaten from its appearance, its aroma and bouquet, its texture and its taste, and the results add up to flavour appreciation. For instance, to appreciate lemon flavour, its taste should be sharp and acid, its aroma should be of lemon, and in some way its appearance must be of lemon—the cream should be lemon coloured, or the biscuit face should resemble or suggest lemon. The texture of the product would condition the appreciation. If the texture is moist (cream or mallow) the flavour will be more obvious. Other conditioning factors are the blend of ingredients and other flavours in the background. One flavour should be dominant and definite, but not too strong. The pH of the product should generally be slightly acid, except for chocolate which shows to best advantage in a slightly alkaline medium. The openness or closeness, brittleness or shortness, sweetness or savouriness of the texture are all factors which further contribute to flavour appreciation. Bearing these factors in mind, and the fact that certain flavours complement each other, such as orange with chocolate, lemon with ginger and so on, then it should be possible to produce biscuits of all types with the optimum flavour appreciation.

CHAPTER 10

Colouring materials

THE use of colours in foodstuffs is almost as important as the use of flavours, and the two are closely associated with each other. Colour should be used to enhance the eye appeal of the product, and the use of correct colour to suggest flavour should be linked directly to the type of flavour in use. Colour should be used at all times with discretion, and it is better to use no colour than too much. If a product has ample materials present to give colour there is no point in adding it, but in most biscuits this is not so, and a small amount of added colour is beneficial.

In considering the uses of colours in food it must be borne in mind that they are used:

(1) To supplement deficiencies of colour. For instance, yellow colour used in a dough was originally introduced to conceal the lack of butter and eggs.
(2) To increase the eye appeal and to complement a definite flavour; and
(3) To introduce varieties and add interest to decorated products.

When colour is used for eye appeal and as a decoration, it is important to use bright clean colours which are not too strong (preferably pastel tints and colours that are associated with food, and with being eaten). In these cases lemon, orange, pink coffee, and chocolate shades are acceptable and popular; green, blue, and purple shades should be used with caution, except when used with a specific and associated flavour. If used in very small quantities, strong contrasting colours can be attractive in conjunction with pastel tints used as decoration.

There are several requirements which a colour must fulfil before it is acceptable for use in foodstuffs, the most important being that it must be harmless. This is not easy to prove, but research is being carried out on this problem and as a consequence of the Colouring Matter in Food Regulations (1957), certain colours are proscribed and others are permitted, but even these are subject to revision. The other requirements of a colour are: it must be readily soluble in water and produce a bright clean colour; it must be unaffected by the action of acids, alkalis, sulphur dioxide, high temperatures, and daylight.

These are very exacting conditions and few colours stand up to them entirely, and as all these factors may occur in biscuit manufacture, care should be used in selecting colours for specific purposes. Reputable manufacturers of colours will supply details of fastness of their products in such conditions.

Although many colours are available, it is necessary to have only the three primary colours, red, yellow, and blue to produce all the other colours and shades. If the primary colours are mixed, they produce secondary colours, thus: yellow with blue forms green; yellow with red forms orange; red with blue forms violet; yellow with red and blue forms chocolate or brown. The shades of colours depend upon the proportion of one to the other, and by experimenting, a complete range of colours and shades can be formed. Before using in production, all colours should be tested in the articles for which they are intended, to check their reactions under the previously mentioned conditions and also under long and adverse storage conditions. Colours should always be examined in daylight, as artificial, particularly fluorescent lighting, can give considerable colour distortion.

The colouring materials that are available fall into two groups: natural and artificial. The natural colours are those obtained from animal and vegetable sources, whereas the artificial colours are aniline dyes obtained from coal tar.

NATURAL COLOURS

Cochineal or carmine

Cochineal is a red from which many pinks are derived. It is prepared from an insect which lives on a variety of cactus originating in Mexico and cultivated in the Canary Isles. The dried insects are powdered and then boiled to extract the colour. The filtered liquid is known as cochineal. Carmine can be precipitated from this liquid by the addition of acid. The precipitate is dried and is available as a powder. To reconstitute, it is necessary to dissolve in a solution of ammonia in water.

By varying the processing treatment of the insect, two forms are available: black-grain and silver-grain. When preparing colours from the insect stage, the ground-up and simmered cochineal colour requires 'fixing' by the addition of alum and lime-water.

Saffron

Saffron is a yellow to orange colour prepared from the stigmas of a crocus grown in Southern Europe and, to some extent, in Cornwall. The dried stigmas are available in cake forms or loose, as hay-saffron. The colour and flavour of saffron is extracted by making a hot water infusion which should be prepared when required, since the colour soon decomposes.

Turmeric

Turmeric is prepared from the dried and ground root of a ginger family plant grown in India and China. Its colour closely resembles egg, and it is sometimes used in egg colours. It has an acrid flavour, but this is not noticeable when it is used in curry powder and mustard pickles.

Annatto

Annatto is a yellow colour, prepared from the fermented fruits of a plant grown in the West Indies and Ceylon. The colour is hardly soluble in water, but is readily so in alcohol. It is used chiefly for colouring butter, cheese, and margarine.

Chlorophyll

The green colouring matter chlorophyll is extracted from the leaves of spinach, nettles, and alfalfa.

Caramel

Caramel or Blackjack, is a dark brown colour prepared by heating sugar until it is decomposing, and then adding boiling water (with care) to form a thick syrup. The use of caramel as a colouring material is safe, and supplies of caramel of a consistent strength can be purchased. It will not produce a chocolate colour, only a brown of marked coffee shade. Caramel will also impart a distinctive flavour of burnt sugar when used in fairly large quantities.

ARTIFICIAL COLOURS

The use of natural colours has largely been superseded since the advent of aniline dyes, as these are generally superior to the natural dyes in reliability, purity, and brightness of colour. Many of the dyes are no longer permitted in foodstuffs. Others are suspect, and even though in use at the present time, may be prohibited in the future. This leads to many problems and difficulties, particularly if exporting from the United Kingdom. Many overseas countries have lists of permitted colours, and probably the shortest and most rigorous list is that of the USA. Where the biscuit manufacturer trades both in the home and export market, it is usual to use the colours permitted by the USA, to avoid duplication and complication. This course is not as straightforward as it might seem, however, because the Brilliant Blue (Blue No.1) permitted by the USA, is proscribed in the United Kingdom. Consequently, chocolate brown colours and greens, suitable for export, are not suitable for home consumption if they contain brilliant blue. This means separate colours and separate production runs, whenever brown and green are in use.

The use of USA permitted colours also increases production costs. The Federal Food, Drug and Cosmetic Act (1938) demands that samples of the colours used in imported foodstuffs should be sent to the USA Food and Drug Administration for analysis, approval, and certification. Although the manufacturers of the colours perform this duty, the customer eventually pays, as these colours are more expensive than the standard colours. Upon certification of the sample, the Federal Authorities issue a 'lot number' which the manufacturers must then quote on all their deliveries to the customers. The ten permitted colours under the Color Additive Amendment of 1960 are usually referred to as F.D. & C. (Food, Drug, and Cosmetic) colours. These colours are red, yellow, green, blue, and violet, and the different shades are numbered thus: 'F.D. & C. Yellow No. 5'. All F.D. & C. colours are permitted for use in Canada and Mexico.

Tartrazine

Tartrazine is a bright lemon yellow of extremely stable qualities. In colour it corresponds to F.D. & C. Yellow No. 5.

Tartrazine is also permitted in European Common Market (EEC) countries, and in the European Free Trade Area (EFTA) countries.

Sunset yellow

Sunset yellow is a good orange colour that is fairly stable. The colour corresponds to F.D. & C. Yellow No. 6 and is permitted in EEC countries, and EFTA countries (Portugal excepted).

Amaranth

Amaranth is red with a definite purple shade. It could be described as deep raspberry red. It is less stable than sunset yellow, but is average taken all round. The colour corresponds to F.D. & C. Red No. 2 and is permitted in EEC and EFTA countries.

Erythrosine

Erythrosine is a bright pink of average stability. In an acid solution it is precipitated owing to its low solubility. It corresponds in colour to F.D. & C. Red No. 3. It is permitted in Denmark, Sweden, and Norway, but not in EEC countries.

Brilliant Blue (F.D. & C. Blue No. 1)

Brilliant blue is not permitted in EEC or EFTA countries (including the United Kingdom). It is extremely stable, and is mainly used in the production of browns, greens, and violets.

Indigo carmine

Indigo carmine is a dark cornflower colour. Its stability is poor. It may be used in the production of browns, greens, and violets, but rather 'muddy' colours will result when compared with those using Brilliant Blue. Indigo carmine corresponds with

COLOURING MATERIALS 99

F.D. & C. Blue No. 2 which is permitted in the United Kingdom, the USA and the EEC and EFTA countries.

Green S

Despite its name, Green S is blue-green in colour. It is permitted in the United Kingdom, Denmark, and EEC countries, and may therefore be helpful in replacing the lack of choice where blue colour is required. It gives a more brilliant colour than indigo carmine and is more stable to baking conditions, but not much more to light.

N.B. Legislation regarding colouring matter in foods is frequently being revised and even where agreement on colours has been reached, for example, in EEC countries, it is often the case that member states still use national regulations, which add to the problems. It is understood that while West Germany has agreed to the common EEC list, no added colouring matter may be present in the crumb of baked goods, only in sugar coatings. Prior to exporting goods to any countries, it is essential that the regulations are thoroughly checked with the agents in the importing countries.

Many other shades are available in the United Kingdom and other countries, but those already mentioned should be sufficient to produce all the necessary intermediate colours and shades. If, for instance, a small proportion of sunset yellow is added to tartrazine, a very satisfactory egg colour will result. All powdered colours should be reconstituted according to the manufacturer's instructions. Water soluble colours should be carefully weighed and whisked in to the correctly measured quantity of boiling water.

To incorporate colours in doughs and batters, the colour should be dispersed in a proportion of the water, preferably before any flour is added. In creams there is a great danger of colours dissolved in water not emulsifying properly, causing the cream to have specks of strong colour in a pale or uncoloured cream. This can be avoided in two ways. The first is to ensure complete emulsification of the colour in a small proportion of the fat, using lecithin as an emulsifying agent before adding to the bulk of the fat. The second is to use proprietary 'non-speck' colours. These are dry colours thoroughly dispersed in a cornflour base. The 'non-speck' colour should be blended with all, or a proportion, of the icing sugar before adding the fat.

PART II
Classification and methods

CHAPTER 11
Classification of biscuit types and methods of production

BISCUITS are broadly classified as being of hard dough or soft dough origin. The hard dough group are savoury, unsweetened, or semi-sweet, and include all types of crackers, puff dough biscuits, and the semi-sweet varieties such as Marie, Rich Tea, and Petit Beurre. In addition to having a low sugar content, or none at all, the fat content rarely exceeds 22·0% of the flour content, except in the case of puff doughs (but even these have a very low fat content at the mixing stage). The soft dough group includes all the sweet biscuits, whether they are plain biscuits, shells, or flow type such as gingernuts. Soft dough biscuits all have many factors in common, but hard dough biscuits fall naturally into three sections: fermented doughs, puff doughs, and the semi-sweet doughs.

THE PRODUCTION OF SOFT DOUGH BISCUITS

Soft or weak wheats, such as English or French, are used in the production of a flour suitable for the manufacturer of sweet biscuits. The flour should have a gluten content in the 7·0-9·0% range, but in certain cases it may be necessary to weaken the flour with cornflour, or boost the structure-forming gluten content by the addition of a proportion of strong flour. The structure of a sweet biscuit is dependent, in the dough form, on the fat and on the partial development of some of the gluten. When baked, the structure is dependent on the coagulated gluten and the gelatinised starch, in conjunction with, to a minor extent, the fat and sugar content. Sweet biscuits of a lean nature, i.e. with a low fat, depend for structure formation on further gluten development, or an increase in the gluten quantity. In either case, there is considerable danger of toughness occurring in the dough, but this is frequently offset by an increase in the sugar content. This has a softening action on the gluten, rendering it more extensible, and consequently, reducing shrinkage caused by toughening. It may be noticed that many biscuits rich in fat have a correspondingly low sugar content and vice versa.

An average quality sweet biscuit has a fat content of approximately 30·0% of the flour content, rising to 35·0% for rich biscuits such as shortcake, and even as high as 45·0% for shortbread. The average quality sweet biscuit will have a combined sugar solids content slightly higher than the fat content, being approximately equal at the 35·0% level, and the sugar content remaining at 35·0%, or even falling to 30·0%, as the fat content increases. At the other end of the scale, as the fat content falls towards 20·0%, the sugar solids content may increase to 45·0%. (All percentages are expressed as percentages of flour content.)

Many other factors, apart from degree of sweetness, affect the sugar content (see Chapter 3, 'Sweetening agents'). Although sugar content is related to the flour content, additional ingredients such as cornflour, oats, or coconut, in sufficient amounts, will require sweetening. If the biscuit is to be sprinkled with sugar, or is to have a sweet filling, or icing, then it may not be so desirable to have a high sugar content in the biscuit itself. The amount and type of sugar is largely responsible for the amount of spread during baking, and for the openness or closeness of the texture, and the shortness or brittleness of the bite.

Methods of mixing: There are two basic methods of mixing soft doughs, but each may have variations designed to achieve the best results under the particular circumstances prevailing, and depending upon the equipment in use.

The first method is known as the 'creaming method', and is probably the traditional way of mixing doughs. It consists of blending the fat and sugars (including syrups and other general ingredients) together, to form a smooth homogeneous cream. During this stage the colours and essences are added. It is unwise to add the colours and essences before the ingredients are becoming blended, as there is a danger of liquid colour being absorbed by one of the ingredients and causing spots of colour to remain in the mixture. While the cream is being prepared, the aerating chemicals and salt are dissolved in some of the dough water. These are added, along with reconstituted milk, to the cream and are mixed in for a short period. The flour is fed into the machine and, as it starts to blend in with the cream, any remaining water is added. Mixing continues until the dough is

CLASSIFICATION OF BISCUIT TYPES

sufficiently developed to suit the type of biscuit-shaping machine for which it is intended.

The second method is known as the 'all-in method', in which all the ingredients are fed into the machine and, as blending commences, the dissolved aerating chemicals, salt, colour and essence, and water are added. This method is more direct and straightforward, but is inclined to produce dough rather more dense and tough than one produced by the creaming method.

Mixing times will vary not only from one type of dough to another, but also from one type of machine to another, but to achieve consistent results it is important that strict attention is paid not only to accurate weighing, measuring, and checking of ingredients, but also to the timing of the various processes and to the temperatures, particularly of the finished dough. This should be 21°C (70°F), but will be influenced by the type of fat in use, the type of shaping unit, and the ambient temperatures.

To run on cutting machines, the mixed dough should be sufficiently developed to be capable of forming a sheet when forced between the sheeting rollers, that is sufficiently strong to withstand subsequent reduction through gauge rollers and to hang together as a network when lifted as scrap from the cut-out biscuits. It should not be moist enough to stick to the rollers, nor dry enough to crack during machining. The scrap should blend with fresh dough to form a good clear sheet. A dough suitable for a rotary moulding requires to be sufficiently developed to form a biscuit shape when under pressure, but needs to be of a drier, more crumbly nature, than that for the cutting machine. If the dough is too dry and crumbly, it will block up the design in the moulding shells. If the dough is too soft, badly shaped biscuits will result. They will be wedged and some will be incomplete.

THE PRODUCTION OF HARD DOUGH BISCUITS

Fermented doughs

Fermentation is a process of decomposition. When a dough made from flour is fermented, decomposition occurs, whereby complex carbohydrates are converted into simple forms and finally to alcohol and carbon dioxide. This process is the responsibility and function of yeast.

Yeast *(saccharomyces cerevisiae)* is a unicellular living organism, almost round in shape, and resembling an egg in construction. The eggshell and membrane are represented by a double cell wall, the egg white by a watery protein substance called protoplasm, the yoke by vacuoles, and the embryo by the nucleus. The entire cell measures only 1/2500th to 1/4000th in in diameter and is composed of 73% to 75% moisture (see Fig. 3). For life and reproduction, yeast requires warmth

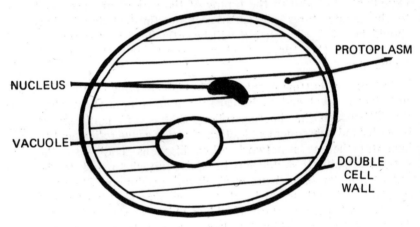

Figure 3. Sketch of a yeast Cell

(optimum temperature of 32°C (90°F)); a food supply consisting chiefly of sugars, but also proteins, minerals salts, and a trace of fat; moisture and air or oxygen. An ideal medium would be a 10-12% sugar solution with dissolved protein and mineral salts, that is slightly acid in nature. Given these conditions, yeast would thrive and commence reproduction. Yeast can reproduce in two ways. One is by budding, and the second is by sporulation.

Budding: Under suitable conditions, reproduction is by budding. The cell begins to swell at one place and a bud gradually appears (see Fig. 4). This bud gradually swells as it is

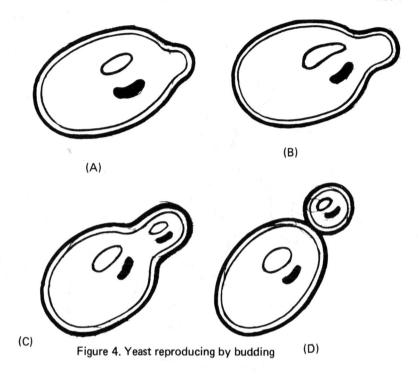

Figure 4. Yeast reproducing by budding

filled with protoplasm from the parent cell, and finally the nucleus and vacuole divide into two, one for each cell. When the new cell is complete, it is detached from the parent and is capable of reproduction itself. This is the process that is required during fermentation of a dough, although in a dough, the medium is not sufficiently fluid, so the yeast cells tend to build up chains or colonies (see Fig. 5).

Sporulation: Under adverse conditions the yeast cell reproduces by sporulation. In this process the nucleus splits into four separate spores and the remainder of the yeast cell dries up, forming a tough protective covering rather like a seed pod covering four seeds. Later, the pod bursts, permitting the spores to circulate freely in the atmosphere. When they come in contact with ideal conditions they will start up fermentation and reproduce by budding. It is. this method of airborne contamination which sets up many uncontrolled and undesirable fermentations.

Figure 5. Colony of yeast cells

Storage of Yeast

Yeast in its compressed form has not very good keeping qualities, except when stored in cool, dry, freely moving atmospheric conditions. If stored in a refrigerator at a temperature of 2-4°C (36-40°F), with air space between the blocks, yeast will keep satisfactorily for 2-3 weeks. In warm, humid conditions, yeast quickly deteriorates. Yeast is obtainable in dried form, but loses some of its activity during the drying process. Approximately half the normal weight of compressed yeast is replaced with dried yeast to obtain similar results. It will, however, keep for considerable periods. The best results are obtained if dried yeast is soaked in a dilute sugar solution for about half an hour before use.

To obtain the best results from yeast when used in a dough, it should be dispersed in a proportion of the dough water at 32°C (90°F), with any yeast food that is to be used in the dough, such as malt products. The yeast must not come in direct contact with the salt, fat, high concentrations of sugar, or alkalis such as sodium bicarbonate. Although its optimum working temperature ranges from 27-32°C (80-90°F), yeast will work at lower temperatures. But, as the temperature decreases, so does the activity, and at temperatures above 32°C the rate of activity increases, becoming uncontrollable until the yeast is destroyed at temperatures above 52·8°C (127°F). At temperatures in excess of 32°C, there are also greater dangers of excess acid fermentation. Although freezing is certainly detrimental to yeast activity, and is not to be recommended, most varieties of baker's yeast used in the United Kingdom will withstand deep freezing ($-17\cdot8$°C (0°F)) for a number of weeks.

Panary fermentation

Fermentation of a dough made from flour is known as 'panary fermentation', and is brought about by the action of enzymes present in the yeast and flour, and to a minor extent by acid bacteria present in the flour. The main action is a breakdown of carbohydrates into carbon dioxide and ethyl alcohol. This breakdown is the result of teamwork between a number of flour and yeast enzymes. Enzymes are often referred to as being 'organic catalysts', which means that they are organic compounds, capable of accelerating or retarding chemical reactions, without themselves being changed in any way by the reaction. They have no life, but are destroyed by heat in excess of 65°C (150°F), and their efficiency is dependent on being in a solution of the correct pH and temperature. In panary fermention the 'available' starch of the flour, i.e. the starch of cells ruptured by the milling process, or by the commencement of germination, or in some cases by the action of the enzyme cytase on the cellulose cell walls of the starch, is then broken down by the action of the group enzyme diastase to form maltose. Both cytase and diastase are flour enzymes. Diastase is also associated with malt products. Diastase is a complex or group enzyme and is often called α (alpha)- and β (beta)-amylase. The breakdown of starch is considered to proceed in two stages: (1) the available starch is converted to dextrins and possibly some maltose by the α amylase, and (2) the dextrins to maltose mainly by the β amylase (see Fig. 6). Maltose is converted to glucose (dextrose) by the action of maltase, sucrose is inverted by invertase to invert sugar, a combination of glucose and fructose (laevulose). Glucose and fructose are broken down by the zymase complex into carbon dioxide and ethyl alcohol. Soluble proteins, such as albumen and globulin of the flour, are attached by protease to form amino acids, peptones, and polypeptides; and lipase attacks the fats to form glycerine and fatty acids. Maltase, invertase, zymase, protease, and lipase, are all yeast enzymes, the first three being significant in their actions, the remaining two playing only minor roles. If germination has commenced in the wheat before milling, the flour will contain many damaged starch cells and also very active proteolytic enzymes, which will soften or even render soluble the gluten upon which the dough structure depends. Some softening of the gluten is necessary, but excessive softening will result in a dough that will not hold together, nor will it withstand machining.

Apart from the production of acid already described, the main source of acid development is caused by the action of lactic acid bacteria (present in flour) upon invert sugar to form lactic acid. About 90% of the acid content of a fermented dough is lactic acid. The lactic acid bacteria prefers a working temperature of 35°C (95°F), and consequently, in short process doughs, there is little, if any, lactic acid produced. Only in

CLASSIFICATION OF BISCUIT TYPES

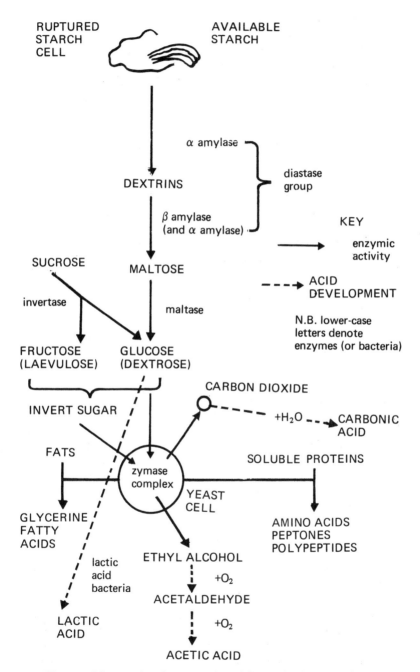

Figure 6. Diagram showing enzymic activity during fermentation

doughs of 3 hr standing is the acid content appreciable, and in overnight sponges or doughs it is considerable. Acetic acid accounts for approximately 5·0% of the acid content, and forms in the later stages of fermentation by the two stage oxidation of ethyl alcohol, first into acetaldehyde and then into acetic acid. The remaining 5·0% is made up of the amino and fatty acids already mentioned; by succinic acid resulting from the action of the zymase complex on invert sugar, and by some of the carbon dioxide formed by fermentation dissolving to form carbonic acid.

Results of fermentation

The results of fermentation should be a softening and mellowing effect on the gluten, rendering it extensible and suitable for machining, and the production of flavour. The improvement of flavour is marked and depends upon the production during fermentation of alcohol, acids, esters, aldehydes, and other trace by-products. The texture, appearance, and eating qualities of the baked goods should be considerably enhanced by correct fermentation.

Over-fermentation occurs when excessive acids are formed and the gluten becomes over-softened even to being corroded and broken down. The dough will be difficult if not almost impossible to handle, and will smell very strongly of acid, vinegary even. In extreme cases of over-fermentation and high temperature it is possible that the lactic acid will be attacked and broken down by butyric acid bacteria to form butyric acid which has a particularly nauseating smell and taste, rendering the dough unfit for use.

In the production of fermented doughs, flours with a fairly high gluten content (10-12%) and strong in nature are used. A strong gluten network is necessary to form structure in the doughs and to withstand the amount of handling in the form of sheeting and lapping which these types of doughs usually undergo. Fermentation then renders the strong gluten extensible, otherwise there would be excessive shrinkage by the biscuits after being cut. The length of time involved in fermentation governs the flour strength. Generally speaking, the longer the process, the greater the flavour development, and to withstand the increase in softening, the flour must be stronger and have a higher gluten content. For short processes a softer flour is necessary, and it will probably be necessary to blend some soft dough flour with the strong flour. Other ingredients in fermented doughs are usually used to enhance flavour, or as yeast foods, or both. Fat is used to increase the extensibility of the gluten and facilitate sheeting, and is used in layers in cream crackers for aeration as already discussed.

The Chorleywood Breadmaking Process

As a result of intensive research into breadmaking processes, the British Baking Industries Research Association (now the Flour Milling and Baking Research Association) developed a revolutionary system known as the Chorleywood Breadmaking Process. Briefly, this system depends upon the use of intense mechanical development in a short space of time. The amount of mixing is recorded by means of an energy input meter in Watt hours. The quantity of energy required is 5 watt hours per pound of dough and this must be expended within a 5-min mixing period. Consequently, very powerful motors are required to drive the mixing machines, and a dough temperature rise of approximately 14°C (25°F) can be expected. The inclusion of fat in low proportions (about 1·0% on flour weight) and an oxidising agent such as ascorbic acid (75 ppm) are necessary in the formula. This intense mechanical development is capable of modifying the gluten structure in a remarkable way, and permits bread of high quality to be manufactured without requiring any period of bulk fermentation. A similar process has been adapted and patented by the Research Association, suitable for the production of cream crackers. The main difference is, that a short period of fermentation (about 30 min) is desirable to obtain optimum results. Cream crackers are being produced commercially by this process.

The Activated Dough Development Process

Further research work on no bulk fermentation time for bread doughs has led to the Activated Dough Development process. This sytem depends upon the inclusion of oxidising agents such as ascorbic acid and potassium bromate, and a reducing agent L—cysteine hydrochloride in a conventional formula and process. No special equipment is necessary, and bread of acceptable standards can be produced without the benefit of any bulk fermentation time. It must be noted, however, that cysteine is not included in the permitted list of improvers of the Bread and Flour Regulations of England, Wales, Scotland, and Northern Ireland.

METHODS OF MIXING

There are three basic methods (not including the method outlined in the previous paragraphs) of mixing fermented doughs, and they are known as the straight dough process; the sponge and dough process; and the flying sponge and dough process.

Straight dough process

The straight dough process consists of dispersing the yeast in a proportion of the dough water along with the malt extract or other yeast foods. The salt and any colour or other chemicals are dissolved in yet another quantity of water. The flour, fat, and remaining dough water, are put in the machine, which is then started. As mixing commences, the salt solution should be added, and then finally the yeast water. The water content and temperature should be correct to produce a dough at the desired temperature and consistency. The dough temperature depends upon yeast content and on the time the dough is to stand fermenting. A secondary factor is the degree of control of dough room temperature and humidity. For doughs that are to stand only a short period in a dough room with a carefully controlled temperature and humidity, a dough temperature up to 32°C (90°F), is in order. For a dough that is to stand 8 hr a temperature of 27°C (80°F), should be quite high enough, and doughs standing longer should be at a lower temperature when mixed. During the period of fermentation, heat is produced and the dough temperature gradually increases. It is important that the dough room temperature be approximately 1°C (2°F) higher than the dough temperature. If there is no control over dough room conditions, the dough should be mixed at a lower temperature than that normally prevailing in the dough room. If

CLASSIFICATION OF BISCUIT TYPES

a dough becomes chilled, owing to a change in dough room conditions, the rate of fermentation will be seriously retarded, and the dough will not be correctly fermented at the time it is supposed to be ready for machining. The dough room relative humidity should be maintained at 80·0% to prevent a tough dry skin forming on the dough. If there is no humidity control, the doughs must be kept covered. Plastic or polythene sheets prevent moisture loss from the dough. If cloths are used, they should be stretched across the dough vessels, leaving a space between the cloths and the dough. This space helps to prevent both heat and moisture loss.

Sponge and dough process

The sponge and dough process consists of making a stiff dough using either 25·0% or 50·0% of the total flour, a corresponding amount of water and either all the yeast or only part of the yeast. This sponge, as it is called, is left to ferment for at least 8 hr and possibly up to 16 hr. During this time, the flour is well and truly fermented, the gluten is considerably modified and, most important, a good deal of flavour develops. When the sponge has stood the allotted time, the remaining ingredients are added as in the straight dough process, and the dough is mixed. Owing to the gluten modification in the sponge, less water will be required in the complete mix. A short final dough fermentation time is sufficient to attain the desired maturity. A very strong flour can be used in the sponge and a soft flour in the dough, and an excellent flavour will result in the finished biscuit. For sponges of this nature, the temperature should not exceed 21°C (70°F), when mixed. Dough room conditions are just as important for this process as for the straight dough process.

Flying sponge and dough process

The flying sponge and dough process consists of preparing a batter composed of approximately one-third of the total dough water with about half this weight of flour. The yeast and yeast food are dispersed in the water, the flour is added, and made into a smooth batter at a temperature of 32°C (90°F). Fermentation ensues rapidly, and the level of the batter rises quickly, making it necessary for the ferment to stand in a correspondingly large vessel. The flying sponge stands for only a period of 20-30 min, during which time the yeast becomes extremely active, even to the point of reproduction, and consequently ensures rapid and thorough fermentation when made into a dough with the remaining ingredients. The flying sponge is intended to give a flying start to fermentation and consequently short time dough processes normally follow. This in turn means that a high yeast content is necessary, and also that there is little development of flavour by fermentation. Flavour has to be supplemented by the use of milk powder, malt products, and even of margarine. On a short process, it is usually necessary to use a higher fat content than with a long process, in order to achieve a greater degree of gluten lubrication as a substitute for the lesser degree of gluten modification by fermentation. For a short process of this nature, it may also be necessary to blend soft flour with the strong flour.

Cream cracker doughs

Fermented doughs are used in the production of cream crackers and also for water biscuits, although water biscuits are also made without the benefits of fermentation, and depend upon enriching ingredients for flavouring and extensibility. Fermented doughs are only suited to cutting machines, and generally require previous treatment on reversing brakes or laminators, before being passed through the gauge rollers.

The basic treatment of cream cracker doughs is to prepare a sheet of dough using the returned scrap as a base and adding 40-50 lb dough on top. When the dough sheet is a suitable thickness, a powder of flour and fat is sprinkled on and the dough is folded over to form a number of laps. Normally, the first stage is composed of 3 layers of dough, enclosing and segregating 2 layers of dust. It is important that the dust is spread evenly and thoroughly, as unevenness at this stage will show up in the baked biscuit as poor quality biscuits or as waste. The dough sheet is reversed, so that the direction of stretch is transverse to the direction in which the dough sheet is then reduced to a suitable thickness for folding again. The number of folds at this stage will vary according to the

CLASSIFICATION OF BISCUIT TYPES

manufacturer's requirements. A 3-layered fold may be sufficient, or folding up to 6 layers may be beneficial. After folding, the dough is sheeted to a thickness suitable for feeding to the gauge rolls of the cutting machine. When reduced to the required thickness, the dough sheet is cut into rectangles which, when reversed in direction, and after they have shrunk, will be of the correct width to fit the gauge rolls. The dough sheets are fed into the machine one after the other with a minimum of overlap, and are joined together by the pressure of the rollers.

When automatic laminators are in use, dough is fed from two separate hoppers in continuous sheets and the layering dust is sprinkled in between by machine. The 2 layers of dough, with the dust between, are then brought together as one sheet by pressure and then lapped. The laminator is at right angles to the cutting machine, so that the sheet of dough is fed from the laminator on to a moving band travelling towards the gauge rolls. The rate at which dough is fed from the laminator is faster than the band moves it away, and by means of a reciprocating delivery feed, the dough is fed back and forth across the moving band, causing a diagonal build up of dough layers. By adjusting the speed of feed, the number of layers in any one part of the final dough sheet can be adjusted, usually approximately 6 layers are standard. The laminated dough sheet is reduced to thickness through sets of gauge rollers, presenting a smooth even surface at the cutter.

Water biscuits are handled in a similar manner but, more often than not, dust is not used in between the layers.

THE PRODUCTION OF PUFF DOUGH BISCUITS

For the production of puff doughs, very strong flour is normally used, as the dough must have a good gluten structure to withstand considerable handling in the form of lapping. The dough generally consists only of flour, water, and salt, mixed to a very stiff dough. The salt content is usually from 2·0-3·0% of the flour content, depending upon the quantity of salt contained in the fat used for layering. Additional ingredients may include fat up to 5·0% of the flour content, and malt extract for flavour.

The inclusion of fat in the dough will facilitate sheeting, but will reduce the available quantity to be used for aeration, and thereby reduce the flakiness of the product. As the dough is not fermented, the baked biscuit relies on the flour, salt, malt if used, and fat for flavour. To achieve the best flavour, butter or a proportion of butter, should be used. Unfortunately, butter becomes very soft during handling, and presents many problems. So it is usual to use speciality fats manufactured specifically for use in puff doughs.

Puff pastry margarine has already been discussed and, although it is admirable from the production point of view, its flavour is marred by the waxy sensation remaining in the mouth caused by the use of high melting point stearins in its manufacture. The puff pastry fats of a similar texture, without the waxy aftertaste, are expensive, and the remaining alternative is the use of fats such as hardened coconut oil or hardened palm kernel oil which have a sharp melting point below blood temperature, thus leaving no waxiness in the mouth. But they are brittle in texture rather than tough and plastic. To be successful, these must be chilled and then minced, and be thoroughly chilled again before distribution in the dough. All temperatures must be kept low, including the dough, dough room, and machine room. Temperatures should not exceed 15°C (60°F). The chilled fat, however, should not be so hard that it breaks through the dough during layering. It should be firm, but should yield under pressure between thumb and forefinger.

The method of mixing a puff dough is quite straightforward. The flour is placed in the machine, the salt is dissolved in some of the water which is then added with the remaining ingredients and water to the flour, to be mixed to a stiff dough. Colour is not often used in puff doughs, but when a yellow margarine is used for layering, the addition of a small quantity of yellow colour to the dough will help to brighten the colour of the interior. This would otherwise incline to look rather dull. It is quite unnecessary to add colour when a white fat is used. When mixed, the dough should be allowed about 20 min to relax in a cool place before layering commences.

The quantity of fat used in puff biscuits varies with the type of biscuit. Biscuits selling on their own merits, such as Butter Puffs, contain 30-35% of the flour content (including that which may have been mixed in the dough), and puff shells for

CLASSIFICATION OF BISCUIT TYPES 119

cream biscuits, which require less lift or aeration and need to be stronger to stand up to mechanical handling on the creaming machine, may contain as little as 15·0%. This means that to 60 lb. dough, approximately 15 lb of fat should be used in the richer-type biscuits. The fat content should be based on the flour content of the dough piece, and as the water content of the dough can vary considerably according to the gluten quantity and quality of the flour, no definite weight can be given. It must also be recalled that margarine may have a water content of up to 16·0% and a salt content of approximately 2·0%, for which allowances must be made.

Handling of a puff dough on the reversing brakes, bears some similarities to those already outlined under cream cracker production. The dough piece is made into a sheet and the chilled minced fat is spread to cover half the dough, leaving the two opposite end quarters uncovered. These are then folded in over the fat to totally enclose it, forming 2 layers of dough with an even layer of fat in the middle. The dough is again sheeted and is given a 3-fold lap, after which it is left to stand for approximately 1 hr. Care should be taken to prevent the dough from forming a dry skin while standing, and it must be kept as cool as possible during this time. This period of standing time is necessary to allow the gluten in the dough to relax. When the dough is handled, the gluten becomes tougher and if all the lapping was applied at once, the gluten would become so tough that it would shrink excessively and become distorted, producing quite useless biscuits. The gluten may even be so over toughened that it would cease to hold together as a network and would break down, resulting in distorted biscuits with uneven or no lift.

When the dough is sufficiently recovered, it should be reversed and sheeted slightly before any scrap is added. The dough is then sheeted sufficiently to enable a 6-fold lap to be given. The direction is reversed and the dough is sheeted to a size suitable for a 3-, 4-, 5-, or 6-fold lap to be given. The number of folds given at this stage will vary according to the quantity and quality of ingredients, the type of biscuit being produced, and the manufacturer's requirements. The dough piece is reversed and reduced to the thickness suitable for feeding to the gauge roll of the cutting machine. It is not usual to cut the sheet in squares, because the more joints there are in the sheet, then the higher is the waste of biscuits, owing to the breakdown of the fat-dough layers, and the biscuits are flat without aeration. The large amount of shrinkage of the dough, after being cut, has to be compensated for by the size and shape of the cutter.

While handling on the brakes, it is important that all the layers are kept as even and intact as possible. The dough sheet should be reduced as gradually as speed of production permits, the fat should be spread evenly, the dough sheet should be kept square at the corners and, when lapped, each fold should be the same size, with no missed spaces and no doubling over. The theory of aeration of puff biscuits has been dealt with (see Chapter 4, Aerating agents), and the lapping process is designed to build up layers of dough with fat in between. As each fold increases the number of layers considerably, at any point where a fold or part of a fold is missed, there will be a deficiency in the number of layers, and anywhere that the fold is doubled over there will be a surfeit of layers. These two faults contribute more than any other to the unevenness of puff biscuits after baking. It must not be assumed that by increasing the number of layers the aeration or lift will be increased. Initially, this is so, but when the optimum number of layers has been passed, the gluten network and the fat are no longer able to remain in their separate layers, and they break down and merge, causing heaviness and unevenness in the baked biscuits. The optimum number of layers will vary according to the type of biscuit and recipe. It will normally be found that the lower the fat content, the lower the number of layers the dough will stand before breaking down. There is, of course, a minimum number of layers (again varying according to the fat content) below which the dough piece will not adhere to itself after cutting and during baking. This results in splitting and flaking, and in some cases the biscuit splits into two separate pieces. This type of fault can also be the result of uneven spreading of the fat or uneven folding.

Whereas gluten relaxation is normally achieved by allowing the dough pieces a recovery period, it is possible to mellow gluten by the use of a proteolytic fungal enzyme. The recovery time should be reduced considerably, or may be cut out completely, if the enzyme is used correctly. Not only can the recovery time be reduced, but the shrinkage after cutting and during baking is not so great. The chief drawback is that the activity is progressive, so that the first and last brake of a mixing will have an appreciable difference in gluten mellowness. This could be overcome by using smaller mixings, so that the standing time of each one is not so great. In the event of breakdowns, the dough must be reduced to a temperature near to freezing point as quickly as possible to retard, if not stop, the enzymic activity.

THE PRODUCTION OF SEMI-SWEET HARD DOUGH BISCUITS

Soft flours are used in the production of semi-sweet, hard dough biscuits, and frequently the flour is weakened by the addition of cornflour, arrowroot, or potato flour. The fat content is relatively low and rarely exceeds 22·0% of the flour weight. The sugar content is normally about 2·0% higher than the fat. The flavour is usually rather bland and is dependent upon milk, syrup, and vanillin, or other added background-type flavours. As the fat content is relatively low and no fermentation is employed to modify the gluten, it might be expected that the resulting biscuits would be hard and tough. In actual fact they are inclined to be tender and brittle. This is achieved by the special mixing technique employed.

The method of mixing is an all-in method, whereby the dissolved salt and aerating chemicals are added to the flour and remaining ingredients, and then mixed until the dough becomes developed as for a fermented dough. Mixing still continues, however, until the gluten becomes softened by the mechanical development and eventually loses its elasticity and is completely extensible. When this stage is reached, threads of dough can be peeled off, with no signs of springing back, but with in fact, a stretch like chewing gum. The mixing time will be two to three times that necessary for a fermented dough, and in a 2-spindle vertical mixer, the mixing time will be approximately 1 hr. During this time there will be considerable heat generated in the dough and the temperature will reach at least 38°C (100°F). After mixing, the dough is allowed to stand for a period of approximately 1 hr. Precautions should be taken to prevent a hard skin forming on the dough, which will readily occur with a dough at so high a temperature.

After this period of relaxation, pieces of dough weighing 50-60 lb are sheeted on the reversing brakes and given a 3-fold lap. No fat or dust is used for layering. The dough piece is reversed, sheeted, and given another 3-fold lap, which may again be repeated. The number of laps given, depends upon results and requirements. The lapping is not intended to give lift or aeration as with cream crackers or puff biscuits, but is to develop structure, a silky smooth surface, and to build a fine, even, bright texture in the baked biscuit. The gluten development is necessary to present a smooth dough at the cutters, which will receive an embossed pattern and will not shrink after being cut.

The method that has been outlined is the basic one, upon which there are variations. The main variation is the addition of an ingredient to hasten gluten development and consequently, reduce the mixing time. Sodium metabisulphite is the usual chemical added in solution to the dough at the mixing stage. This has a quite remarkable mellowing effect on the gluten, which is brought about by sulphur dioxide. The solution deteriorates on storage, so only one day's requirements should be prepared at a time. The quantity of sodium metabisulphite used is normally in the region of 0·05% of the flour content. The use of sodium metabisulphite in doughs has led to the adoption of the 'Melloene' treatment of flour by the millers. (This has already been mentioned under flour treatment.) Similarly, the use of a proteolytic fungal enzyme will accelerate gluten softening, resulting in reduced mixing times and also in a reduction of standing time before braking commences. The quantity to use will again depend upon the type of flour available, and also on the dough temperature and fat and sugar concentrations, and finally, on the activity or strength of the enzyme preparation supplied by the manufacturer. The influence of temperature is one of acceleration as temperature increases; as the fat and sugar concentration increases, enzyme activity is retarded. The enzyme preparation will be of consistent strength, and advice upon this will be readily supplied by the manufacturer. A quantity in the region of 0·025% of the flour will probably be found sufficient. The enzyme should be dispersed in water before addition to the dough, or it can be fed into the flour stream by the miller to the baker's requirements. It must be recalled that enzymic activity is progressive and, in the event of a breakdown, the dough temperature must be reduced to near freezing point as quickly as possible, otherwise the dough will become quite unmachineable. It may be necessary to divide the dough into pieces to ensure rapid temperature reduction, even in a refrigerator. If the dough becomes unusable because of excessive softening, it can be worked away, a constant proportion at a time, in subsequent doughs, with a corresponding reduction or complete absence of the enzyme in the mixing.

The lapping of semi-sweet, hard doughs, can be achieved by the use of laminators in a similar manner to that described for cream crackers, but only one hopper is necessary for the dough feed and no dust is sprinkled.

In order to reduce the labour costs involved in lapping these doughs on reversing brakes, many manufacturers have developed the formulae to be able to feed the dough directly into the hopper of the pre-sheeter and then to the gauge rolls and cutter, without any lapping at all. Quite satisfactory results can be obtained this way, but the fineness and brightness of the biscuit texture are considerably reduced. The main adjustments necessary to the recipe are to the aeration chemicals.

PART III
Formulae-quality control and development

CHAPTER 12

Basic ingredient proportions of biscuit doughs

ALTHOUGH a variety of formulae are to be dealt with, it cannot be over-emphasised that these are intended as a base upon which to build satisfactory products. Owing to the variation in the ingredients used, but also in the handling methods of those ingredients and the doughs, it is not possible to lay down hard and fast recipes that will produce a perfect or, for that matter, an acceptable biscuit at the first attempt. However, it should be possible, from the given formulae and the information on the raw materials used, and the results of laboratory tests and experiments on small batches, to produce a biscuit to satisfy both the manufacturer's and the consumer's requirements. During the tests and experiments a strict record should be kept of all variations of ingredients and techniques, even of mistakes. In this way experience will be gained of the ingredients and equipment used, and once there is a record it can always be used for future reference.

GUIDE TO USING FORMULAE

As the standard weight of flour varies from place to place, the ingredient quantities in the recipes will all be based as percentages of the flour weight. In this way the figures can be expressed as pounds, ounces, or grammes, depending upon the size of mixing required or converted to represent weights against a standard flour weight (see Table 4). All fat contents will be assumed to be 100·0% fat, so that if the fat or a proportion of the fat used is butter or margarine, an allowance must be made for the water content, similarly, an allowance must be made for the salt content, the figures given for glucose, syrup, and malt extract, are based on the dry solids content, so, when converted, an extra proportion must be added to cover a moisture content of approximately 20·0%. The moisture contents of other ingredients are not taken into account.

The use of colours and flavours will be indicated, but these are optional. As qualities vary considerably, the quantity to use must initially be based on the manufacturers' recommendations and then on the biscuit makers' own judgement and require-

TABLE 4: *Table converting ingredient percentages to pounds when based*

%	lb	%	lb	%	lb	%	lb	%	lb	%
100	280·0	90	252·0	80	224·0	70	196·0	60	168·0	50
99	277·2	89	249·2	79	221·2	69	193·2	59	165·2	49
98	274·4	88	246·4	78	218·4	68	190·4	58	162·4	48
97	271·6	87	243·6	77	215·6	67	187·6	57	159·6	47
96	268·8	86	240·8	76	212·8	66	184·8	56	156·8	46
95	266·0	85	238·0	75	210·0	65	182·0	55	154·0	45
94	263·2	84	235·2	74	207·2	64	179·2	54	151·2	44
93	260·4	83	232·4	73	204·4	63	176·4	53	148·4	43
92	257·6	82	229·6	72	201·6	62	173·6	52	145·6	42
91	254·8	81	226·8	71	198·8	61	170·8	51	142·8	41

ments. The term 'VOL' refers to the volatile salt, ammonium bicarbonate, which is also known as 'ammonia'. Unless otherwise specified, the mixing and handling techniques are those discussed in the preceding chapter.

SOFT DOUGH BISCUITS *(a) Rotary moulded and cut biscuits*

SHORTCAKE (three qualities):

	Lean	Average	Rich
Flour, soft	100·0	100·0	100·0
Fat	25·0	30·0	35·0
Sugar, pulverised	35·0	32·0	28·0
Syrups	4·0	2·0	2·0
Milk powder, skim	2·0	2·0 Full cream	1·0
Salt	1·17	1·08	1·0
Sodium bicarbonate	0·36	0·4	0·45
Vol	0·27	0·27	0·18
Cream powder	0·18	0·09	0·09
Essences	Vanilla and butter		
Colour	As required		

Although shortcake is usually rectangular or finger shaped, this is a popular type of biscuit, suitable for different shapes, which may be given topical or local names. A further variation is to sprinkle with sugar before baking. The syrups used in the mixings are optional between glucose and syrup or a combination of the two. Syrup will add extra flavour, but will tend to discolour the crumb. If margarine or butter and syrup are weight of 280 lb (i.e. one sack) and vice versa.

lb	%	lb	%	lb	%	lb	%	lb	Equiv. in oz
112·0	30	84·0	20	56·0	10	28·0	0·5	1·4	–
109·2	29	81·2	19	53·2	9	25·2	0·36	1·0	16
106·4	28	78·4	18	50·4	8	22·4	0·315	0·875	14
103·6	27	75·6	17	47·6	7	19·6	0·27	0·75	12
100·8	26	72·8	16	44·8	6	16·8	0·225	0·625	10
98·0	25	70·0	15	42·0	5	14·0	0·18	0·5	8
95·2	24	67·2	14	39·2	4	11·2	0·135	0·375	6
92·4	23	64·4	13	36·4	3	8·4	0·09	0·25	4
89·6	22	61·6	12	33·6	2	5·6	0·045	0·125	2
86·8	21	58·8	11	30·8	1	2·8	0·0225	0·0625	1

used, it is advisable to use a small proportion of yellow or egg colour, otherwise the crumb tends to be dull.

SHORTBREAD:

Flour, soft	95·0
Cornflour	5·0
Fat	45·0
Sugar, pulverised	36·0
Milk powder, full cream	2·0
Salt	0·9
Sodium bicarbonate	0·18
Vol	0:09
Essences	Butter and/or vanilla or lemon
Colour	As required

Shortbreads are the richest and most expensive biscuits produced, and they should be considered only if they can be made a really economic proposition. The true flavour of shortbread is butter. This cannot be replaced by essences or substitutes, therefore, a good proportion of the fat should be butter (if not all of it). The creaming method of mixing is recommended, as this is superior to the all-in method for aeration, and it will be noticed that the aerator content in the recipe is very low. Shortbread is produced in various shapes and sizes, from 'petticoat tail' to blocks weighing up to a pound. It is usually produced in two thicknesses. The thinner are similar to most other biscuits, but the thicker may be ¾ in (19 mm) thick. These require baking in frames which permit the shortbread to rise, and help to maintain the open texture that is desirable. Rice flour is sometimes used in the dusting flour employed to prevent sticking at the gauge rollers. Some shapes of shortbread may be given a light sprinkling of castor sugar.

LINCOLN (Lincoln cream):

Flour, soft	100·0
Fat	32·0
Sugar, pulverised	33·0
Syrups	2·0
Milk powder, skim	0·9
Salt	1·08
Sodium bicarbonate	0·36
Vol	0·27
Cream powder	0·09
Essences	Vanilla and butter
Colour	As required

Lincoln are a relatively rich biscuit, usually about 2¼ in (58 mm) diameter, and covered in a pattern of tiny hemispheres. They are often called Lincoln cream, but are not used as shells for a cream filling. Special Lincoln cream flavour can be purchased from essence manufacturers, in which case the recommended essences can be replaced by it if preferred.

NICE:

Flour, soft	100·0
Coconut, fine	15·0
Fat	20·0
Sugar, pulverised	28·0
granulated	8·0
Syrups	4·0
Milk powder, skim	3·0
Salt	1·125
Sodium bicarbonate	0·4
Vol	0·27
Cream powder	0·18
Essences	Coconut and butter

Nice are usually a rectangular coconut biscuit, liberally sprinkled with sugar. The texture is normally light and open and the bite tender. Nice are frequently cut in a similar size to a custard cream shell and are sandwiched with a cream filling containing coconut.

SHELLS

SHELLS FOR CREAMING (two qualities)

	Average		Rich
Flour, soft	100·0		100·0
Fat	27·5		30·5
Sugar, pulverised	28·5		28·5
Syrups	3·0		2·0
Milk powder, skim	1·8	Full cream	0·9
Salt	1·08		1·08
Sodium bicarbonate	0·45		0·36
Vol	0·27		0·27
Cream powder	0·09		0·18
Essences	Vanilla or other suitable flavours		
Colour	As required		

Shells suitable for creaming are produced in many shapes and sizes, and for many different flavours of cream filling. Wherever possible, the flavour used in the shell should blend with that of the cream. If a soft, fruit-flavoured cream is to be used, it is usual to choose a shell with only background flavour, such as vanilla or butter, neither of which should be strong or dominant. For a custard cream, the shell has usually a marked vanilla flavour, but it must always be borne in mind that the use of flavourings is more efficient in the filling than in the shell that is baked. The use of strong flavours in the shell should be avoided, except where a definite flavour combination is required, such as a ginger-flavoured shell, combined with a lemon-flavoured cream. In this case, both are strong flavours, but neither should predominate; each should complement the other to form a perfect blend.

CHOCOLATE-FLAVOURED SHELL (Bourbon type):

Flour, soft	100·0
Fat	25·0
Sugar, pulverised	35·0
Syrups	2·0
Milk powder, skim	0·9
*Cocoa powder	4·5
Salt	1·08
Sodium bicarbonate	0·45
Vol	0·27
Cream powder	0·18
Essences	Vanilla and butter
Colour	Chocolate colour

* The cocoa powder content is below the legal limit for the biscuit to be termed a 'chocolate biscuit', hence the use of the phrase 'chocolate flavoured'.

If the chocolate flavour is insufficiently strong, then the cocoa powder could be increased, or chocolate flavour added. To obtain a good bright chocolate colour, it is usually necessary to add extra red colour, either independently, or to the original blend of colour. In the production of Bourbon shells (a finger-shaped shell), it is usual to sprinkle lightly with sugar. Cocoa is not always included in Bourbon shells and, if omitted, extra colour will be necessary (so long as no reference is made to chocolate, then this action will not contravene the regulations). If the resultant shell is satisfactory, by substituting the chocolate colour with a little yellow colour, another plain, sweet biscuit recipe is produced.

BASE FOR MARSHMALLOW:

In the production of marshmallows, it is necessary to use a biscuit base that has attracted moisture from the atmosphere. To accelerate this conditioning period, the biscuit base usually has a high invert syrup content.

Flour, soft	100·0
Fat	20·0
Sugar, pulverised	30·0
Glucose	5·0
Invert syrup	10·0
Salt	1·26
Sodium bicarbonate	0·36
Vol	0·27
Cream powder	0·09
Essences	Vanilla
Colour	As required

Bases for marshmallows can be made in various shapes and sizes to suit the type of mallow product in demand. The base should be coloured and flavoured in accordance with the requirements and the mallow flavour.

DIGESTIVE:

Flour, soft	70·0
Wheatmeal	30·0
Fat	28·8
Sugar, granulated	10·0
fourths (pieces)	10·0
Demerara	6·0
Syrups	4·0
Milk powder, skim	3·0
Salt	1·26
Sodium bicarbonate	1·0
Vol	0·45
Cream powder	0·09
Essence	Butter
Colour	As required

Digestive biscuits contain a wheatmeal or wholemeal flour with a fairly coarse bran content. Some manufacturers supplement the wheatmeal by the addition of clean bran, and it is possible to add broken grains of wheat, oats, or barley, to give the biscuit its characteristic flavour and roughness. When using broken grain, care should be taken to see that it is neither too coarse nor too hard. If it is either, there is a likelihood of machining trouble at the rollers and a tendency for the dough sheet to crack and break. In the baked biscuit, hard pieces of grain can be unpleasant, if not dangerous, when being eaten. It is usual to include a coarse-grained sugar, such as Demerara, to open up the texture and to enhance the flavour. As dark sugars and flour are in use, the syrup content will improve the flavour if golden or manufacturer's syrup is used. Digestive, which are also referred to as 'sweetmeal', are produced either round or oval, and are baked on wire bands as well as steel bands. When baked on a steel band, the biscuit will have considerably more volume than when baked on a steel band. Digestive are a popular line plain, or when half-coated with chocolate.

Fruit biscuit:

Flour, soft	100·0
Cornflour	2·5
Fat	30·0
Sugar, pulverised	16·0
fourths (pieces)	16·0
Syrup	3·6
Milk powder, skim	2·0
Salt	1·08
Sodium bicarbonate	0·54
Vol	0·27
Cream powder	0·27
Essence	Butter, lemon and/or orange
Minced peel	1·8
Currants	12·5

Fruit biscuits have a limited appeal as a line in their own right, but they add a useful variety to assorted packs. They are usually round, but other shapes may be more satisfactory, according to the use for which they are intended. A very attractive finish is achieved by washing over the biscuits with water and then sprinkling with sugar. If some of the sugar dissolves during baking, then the covering breaks up with a mosaic effect, and there should be no loose sugar remaining on the surface. The main disadvantage of fruit biscuits, is that the fruit becomes broken up and discolours the biscuit. This is unavoidable, but can be minimised by using small, good quality fruit, that is quite dry (washing or soaking softens the skins), and by adding the fruit to the mixing at the last possible moment to ensure even distribution. On a cutting machine, the scrap return should be the least possible. This formula should produce a biscuit that is tender and crisp to eat, with a light open texture.

GINGERNUTS:

Flour, soft	100·0
Fat	17·5
Sugar, granulated	10·0
fourths (pieces)	40·0
Demerara	10·0
Syrup	20·0
Salt	1·0
Sodium bicarbonate	1·0
Vol	0·18
Cream powder	0·18
Ginger (ground)	1·35
Essence	Lemon
Colour	As required

Ginger biscuits are traditionally hard and brittle. They spread considerably during baking and should develop a degree of 'crack' on the surface. The hardness of the biscuit can be attributed to the high sugar and syrup content, and to the relatively long baking time which causes partial caramelisation of some of the sugars. The biscuit flow is a result of the high sugar content and to the high sodium bicarbonate content, both of which have considerable softening effect upon the gluten structure. The degree of crack is dependent upon the ratio of sugar and crystal size. If all the sugar is fine in crystal size, the crack will be very fine and the biscuit will spread excessively. To produce an attractive crack, the biscuit should not flow excessively, and some large-grained sugar should be used to break up the surface in controlled fissures. The biscuits should be well washed, and then baked with steam and high humidity in the oven. The wash and oven humidity are other important factors influencing the flow and degree of crack.

The use of lemon flavouring in ginger biscuits is optional, but its presence in small proportions is the usual practice, and it helps to bring out the full flavour of the ginger. Colour is not entirely necessary. Owing to the use of dark sugars, syrup, ginger, and a slow baking speed, sufficient caramelisation should occur to give the biscuit a pleasant shade. The colour will be enriched by the use of an egg colour, or a ginger colour produced by adding a small quantity of sunset yellow, or even less amaranth, to tartrazine. Some manufacturers favour the inclusion of mixed spice in gingers, in which case the addition of 0·045% of the flour weight should be sufficient.

GINGER BISCUITS:

Flour, soft	100·0
Fat	30·0
Sugar, granulated	20·0
fourths (pieces)	20·0
Demerara	10·0
Glucose	4·0
Syrup	10·0
Salt	1·08
Sodium bicarbonate	0·9
Vol	0·45
Cream powder	0·36
Ginger (ground)	1·26
Essence	Lemon
Colour	As required

This biscuit has a considerably reduced sugar and syrups content and increased fat content. The resulting biscuit should spread, and develop crack (probably to a lesser degree), but should not be so hard as the traditional product. In fact, it should be crisp and tender to eat.

FRUIT CRUNCH (flow type):

Flour, soft	100·0
Fat	34·0
Sugar, granulated	36·0
fourths (pieces)	12·0
Syrups	10·0
Milk powder, skim	2·0
Salt	1·08
Sodium bicarbonate	0·72
Vol	0·45
Cream powder	0·36
Essence	Orange or suitable flavour
Colour	As required
Dried fruit	14:0

This type of recipe should be treated in a similar manner to that for gingernuts, in order to produce a hard crunchy biscuit. It should not be as hard as the gingernut, owing to the high fat content. If currants are used, they are probably best used whole. When minced, the skins show up dark and bitty and detract from the appearance of the biscuit. Light-coloured fruit, such as sultanas and raisins, can be minced so that a consistent flavour results throughout the biscuit. The essence should either supplement or complement the fruit in use, or some of the fruit may be replaced by minced peel (e.g. reduce fruit to 12·0%, add minced orange peel–2·0%).

SOFT DOUGH BISCUITS

(b) Wire cut biscuits

BASIC COOKIE:

Flour, soft	100·0
Fat	32·0
Sugar, pulverised	34·0
Syrup	3·0
Milk powder, skim	2·0
Salt	1·08
Sodium bicarbonate	0·45
Vol	0·36
Cream powder	0·27
Essence	Suitable
Colour	As required

This formula is given as a basic one from which different types and flavours can be developed. The texture can be altered by the substitution of coarse sugar for some of the pulverised sugar. Nuts or fruit can be introduced. Any one of the following additions could be made to the basic recipe to produce different lines:

(1)	Nibbed cashew (or other) nuts	15·0
(2)	Currants	15·0
(3)	Nibbed cashew nuts	5·0
	Chocolate nibs	12·5
(4)	Coconut fine (or other ground nuts)	10·0
	Currants	10·0
(5)	Chocolate nibs	10·0
	Currants	10·0
(6)	Coconut, fine (for coconut ring)	20·0

The essences used would vary in each case, but should be suitable to the added ingredients. The use of vanilla or butter flavour could be used as background in conjunction with nut or fruit flavours. Various dies can be used, but it must be recalled that when fruit or chocolate nibs are included in a mixing, they should pass easily through the die. Fruit should be small, of good quality, and dry, and should be added at a late stage of the mixing process.

COCONUT RING (top quality):

Flour, soft	100·0
Coconut, fine	30·0
Fat	28·0
Sugar, granulated	25·0
fourths (pieces)	20·0
Milk powder, skim	2·0
Salt	1·26
Sodium bicarbonate	0·36
Vol	0·27
Cream powder	0·18
Essence	Coconut and butter

This biscuit is a big improvement on the coconut ring variation of the previous recipe. The biscuit will spread more during baking, but will be crisper, sweeter, and more nutty.

SOFT DOUGH BISCUITS

(c) Rout press biscuits

Many of the previous formulae can be adapted to produce satisfactory biscuits on the rout press machine. Various dies can be used in the production of interesting and attractive lines. The following recipe is a typical good quality rout press biscuit.

Flour, soft	100·0
Coconut flour (optional)	5·0
Fat	30·0
Sugar, pulverised	32·0
Syrups	5·0
Milk powder, skim	1·8
Salt	1·08
Sodium bicarbonate	0·45
Vol	0·36
Cream powder	0·18
Essences	Vanilla and butter
Colour	As required

Variations of texture can be achieved by substituting some of the pulverised sugar for a coarser grained sugar, but the choice must be made with consideration of the size of aperture in the die. This also applies when coconut is to be used.

FERMENTED DOUGH AND PUFF DOUGH BISCUITS

The handling and control of fermented and puff doughs has been dealt with in the preceding chapter. It will be recalled that to achieve the desired standard in the produce, flour strength, yeast content, temperatures, and handling methods are the critical factors that will probably require adjusting, rather than the other ingredients. Sodium bicarbonate may be added to neutralise acid development and to facilitate handling properties of the dough sheet (up to 1:0% of the flour weight). This will result in an improvement in the appearance of the crackers also, but will be detrimental to the flavour.

CREAM CRACKERS
(1) Straight dough process

Flour, strong	75·0
soft	25·0
Fat	15·0
Malt extract	2·16
Milk powder, skim	0·9
Yeast	1·35
Salt	1·26
Colour	As required
Dough temp. when mixed	32°C (90°F)
Dough time	3 hr

(2) Sponge and dough process

Sponge

Flour, strong	50·0
Yeast	0·54
Water: To mix to very stiff dough.	
Sponge temp. when mixed	21°C (70°F)
Sponge time	13 hr

Dough

Flour, strong	50·0
Fat	14·0
Malt extract	1·08
Milk powder, skim	0·9
Yeast	0·36
Salt	1·26
Colour	As required
Dough temp. when mixed	32°C (90°F)
Dough time	3 hr

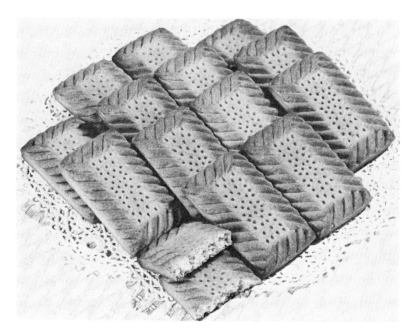

Plate 1a. Shortcake fingers

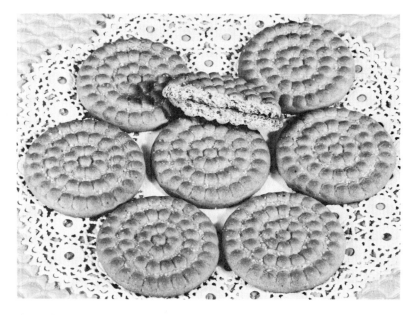

Plate 1b. Round shortcake biscuits

Plate 2a. Shortcake biscuits: 'Royal Duchess'

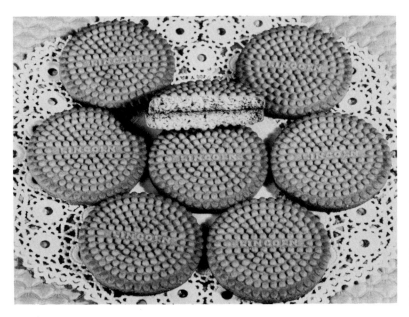

Plate 2b. Lincoln (Lincoln Creams)

Plate 3a. Nice

Plate 3b. Finger creams

Plate 4a. Custard creams

Plate 4b. Bourbon creams

Plate 5a. Digestive (Sweetmeal)

Plate 5b. Currant biscuits

Plate 6a. Gingernuts

Plate 6b. Half-coated (chocolate) sweet biscuits

Plate 7a. Coconut cookies (wire-cut)

Plate 7b. Cookies with chocolate, nuts and currants included

Plate 8. Cream crackers

Plate 9a. Savoury crackers

Plate 9b. Creamed puff shells

Plate 10a. Rich Tea biscuits

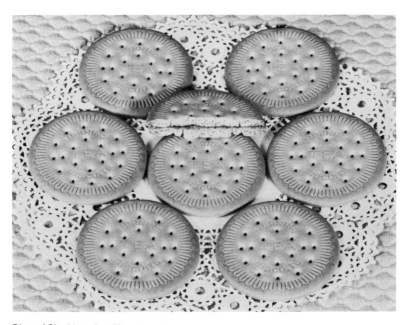

Plate 10b. Morning Tea biscuits

Plate 11a. Marie biscuits

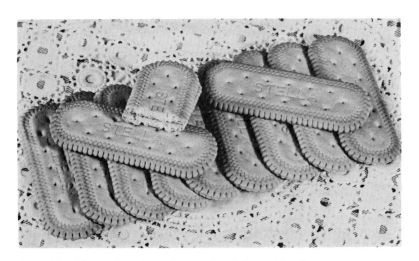

Plate 11b. Finger-shaped, semi-sweet, hard dough biscuits

Plate 12a. Garibaldi biscuits

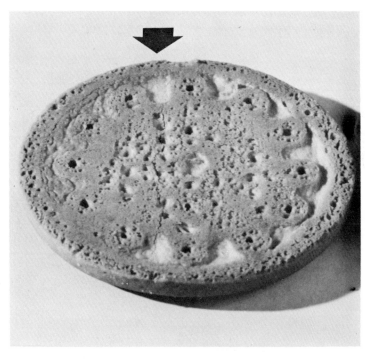

Plate 12b. Morning tea biscuit showing hair-line fracture known as 'checking'

Plate 13a. Coconut mallows

Plate 13b. Chocolate teacakes

Plate 14a. Jam rings

Plate 14b. Fig bars

Plate 15a. Cream filled wafers

Plate 15b. Bag-type pack of coconut cookies

Plate 16a. Tray-type pack for mallows using a preformed liner

Plate 16b. Tray-type pack for assorted biscuits using a preformed liner

BISCUIT DOUGHS

(3) *Flying sponge and dough process*

Flying sponge	Flour, strong	6·0
	Malt extract	2·88
	Yeast	1·8
	Water: Approximately double the flour weight.	
	Sponge temp.	32°C (90°F)
	Sponge time	20-30 min

Dough	Flour, strong	69·0
	soft	25·0
	Fat	16·0
	Milk powder, skim	1·8
	Salt	1·26
	Colour	As required
	Dough temp.	32°C (90°F)
	Dough time	1¾ hr

In each case, approximately 18·0% of the total flour weight of layering dust is used for each dough.

Layering dust:

	Flour, soft	100·0
	Fat	35·72
	Salt	4·64

The dust is made by chilling the fat and by pulping it into the sieved flour and salt. The dust is then mixed until it is a very fine crumb. The best type of fat for this purpose is hardened coconut oil, or hardened palm kernel oil. The use of salt in the dust is optional. Its purpose is to compensate for the additional flour that is to be lapped into the mixing, for which there is no salt in the original dough.

For the methods of handling, reference should be made to the section on fermented doughs. Crackers of all types and puff biscuits are normally baked on wire bands.

Cheese or Savoury Crackers
(Straight dough process)

Flour, strong	75·0
soft	25·0
Fat	14·0
Malt extract	2·88
Milk powder, skim	0·9
Cheese powder	8·0
Yeast	1·08
Salt	1·26
Dried autolysed yeast	2·5
Pepper (ground)	0·09
Essence	Cheese
Colour	As required
Dough temp. when mixed	32°C (90°F)
Dough time	3 hr
Layering dust	9·0

A similar process is used in the manufacture of cheese crackers as for cream crackers. The layering dust is the same, but only half the quantity is normally necessary. The crackers are usually cut with a small, circular, indented cutter, and are lightly sprinkled with salt before baking. After baking, their appearance is enhanced by the application of an edible oil spray. For this purpose, hardened coconut or palm kernel oils are used, but groundnut oil is also suitable. This formula would be satisfactory to produce plain savoury crackers by omitting the cheese, the pepper, and the cheese flavour. Other flavours or spices could also be substituted. The cheese powder referred to in the formula, would have a maximum moisture content of 4·0%. Therefore, if fresh cheese is to be used, it must be increased by a corresponding amount.

Water Biscuits:

	Fermented	*Unfermented*
Flour, soft	100·0	100·0
Fat	9·0	11·0

BISCUIT DOUGHS

WATER BISCUITS (cont.)

Sugar, granulated	3·5	3·5
Syrup	–	1·44
Milk powder	–	1·44
Yeast	1·26	–
Salt	1·26	1·26
Colour	As required	As required
Dough temp.	80°F	70°F
Dough time	2 hr	1 hr

Water biscuits are handled as for cream crackers, but no layering dust is used when the dough is lapped. Both are extremely lean recipes. The fermented one has the benefits of fermentation, whereas the unfermented formula has milk powder and syrup to add to the flavour and extra fat to improve the handling qualities.

PUFF BISCUITS DOUGH

Flour, strong	100·0
Malt extract	2·6
Salt	2·5
Colour	As required

Layering fat—lean biscuits 18·0%; rich biscuits 36·0%.

Part of the fat content may be included in the dough, and when margarine is used, the salt content should be adjusted accordingly. When white fat is used, there is no necessity to add colour to the dough, but it brightens the texture of the biscuits when margarine is used. The temperature of all ingredients, equipment, and departments should be kept as low as possible during the production of puff biscuits.

Puff biscuits should be cut with the minimum of scrap, which is returned to the dough as evenly as possible during the second lapping stage. The cutters should allow for considerable shrinkage in the dough piece after cutting, and where possible it is advantageous to use a scrapless cutter. Puff shells and fruit-filled puff biscuits, are usually sprinkled with sugar before baking. The sugar should be fine enough to melt and partially caramelise during baking, to form an attractive glaze.

SEMI-SWEET HARD DOUGH BISCUITS

The production of semi-sweet, hard dough biscuits, has been dealt with in the preceding chapter, and reference should be made to it, particularly regarding the inclusion of ingredients to hasten gluten softening, which will not appear in the following formulae.

RICH TEA

Flour, soft	100·0
Cornflour	2·5
Fat	20·0
Sugar, pulverised	24·0
Syrups	2·0
Milk powder, skim	3·6
Salt	0·72
Sodium bicarbonate	0·54
Vol	0·54
Essence	Vanilla and butter
Colour	As required

By reducing the sugar content to 20·0%, this recipe would also be suitable for Osborne biscuits.

MARIE

Flour, soft	100·0
Cornflour	4·0
Fat	18·0
Sugar, pulverised	22·3
Syrups	2·5
Milk powder, skim	2·7
Salt	0·72
Sodium bicarbonate	0·54
Vol	0·45
Cream powder	0·18
Essences	Vanilla and butter
Colour	As required

Petit Beurre

Flour, soft	100·0
Cornflour	5·0
Fat	17·2
Sugar, pulverised	20·0
Syrups	3·0
Milk powder, skim	2·7
Salt	0·72
Sodium bicarbonate	0·72
Vol	0·36
Cream powder	0·27
Essences	Butter
Colour	As required

Petit beurre are normally baked on a wire band, preferably of a 'Continental' or 'herring-bone' type.

Arrowroot

Flour, soft	100·0
Cornflour	7·2
Fat	16·0
Sugar, pulverised	19·0
Syrups	4·5
Milk powder, skim	1·8
Salt	0·72
Sodium bicarbonate	0·72
Vol	0·27
Cream powder	0·18
Essences	Vanilla and butter
Colour	As required

Although cornflour is most frequently included, traditionally, arrowroot was used. The name is often prefixed by the word 'thin', as these are usually sheeted very thinly to produce a wafer-like biscuit.

Any of these semi-sweet, hard dough biscuit formulae could be used satisfactorily in the production of shells for creaming.

Small variations may be necessary according to cost, sweetness, and flavour required. Fruit biscuits of the Garibaldi type can be produced by feeding two dough sheets to enclose a sprinkled fruit filling prior to the gauge rollers. Approximately 40·0% of the total flour weight of currants will be necessary for an adequate covering. The double dough sheet, with the fruit between, is reduced to the desired thickness and is cut into squares or rectangles by a scrapless cutter. The biscuits should be given a milk wash to produce a pleasing surface colour and finish.

BAKING OF BISCUITS

Although the baking procedure is probably the most critical process in biscuit manufacture, it is not possible to lay down rules regarding the actual process, because of the many variations which may be involved. These variations include the type of oven in use, the different manufacturers' versions of similar ovens, the method of heating and the type of fuel used. In addition, there are extra devices such as high-frequency or electronic attachments, designed to speed up the baking time. The correct baking conditions are those which will produce a biscuit of the desired appearance and texture, but also of minimum moisture content (approximately 1·0%) and in the interests of low production costs in the minimum time. Even though the moisture content is reduced to a minimum, it must be evenly distributed in the biscuit, otherwise there will be danger of checking. Baking time is the main factor which affects productivity, and it is upon this that all other facets of production must be based. Baking times are chiefly governed by the type and size of the biscuit, but other considerations are the ratio of weight to area, the closeness of the biscuits on the band, and the degree of coverage on the band. A thick finger biscuit will require a slower baking time than a thin flat biscuit of the same weight to ensure complete baking and even moisture distribution. Rectangular biscuits cover more oven band than round biscuits, and, because of their width, some biscuits leave more space between the rows than others, according to the design of the cutter or roller. Bearing these variable factors in mind, the following baking times may be used for guidance:

BISCUIT DOUGHS

SOFT SWEET DOUGHS:	5 min for light shells for creaming.
	8 min for large biscuits
	6½-7 min on average.
Flow type:	7½-10½ min according to size.
Wire cut:	1½-2 min longer than similar weight of sweet biscuit.
SEMI-SWEET HARD DOUGHS:	Similar to soft doughs.
CRACKERS, VARIOUS TYPES:	4-5 min.
PUFF DOUGHS:	9-12 min. according to sizes.

N.B.: Reduced times will be possible in 'turbulence' ovens and those with electronic attachments.

CHAPTER 13

Basic ingredient proportions of wafers, marshmallows, creams, and fillings

WAFERS

WAFERS, as used for sandwiching with a cream filling and for ice cream, are virtually made from a batter of flour and water. They are, however, improved by the addition of enriching ingredients and chemicals for aeration. The following formula is intended as a basic recipe:

Flour, wafer	100·0
Milk powder, skim	2·5
Salt	0·75
Sodium bicarbonate	0·2-0·3
Colour	As required
Water	140·0-160·0

Although wafers can be made using flour of almost any strength, the best results are obtained when medium strength flour is used. This will probably entail blending soft flour with strong until the optimum blend is achieved. A strong blend will require more water to mix to the correct consistency, resulting in a greater yield of wafer sheets per mix, but naturally each sheet will be lighter than those made from a weak flour with low water absorption. This factor is important from a costing point of view, depending upon whether the sheets are sold by weight or by number. Milk powder is not necessary to the formula, but improves the colour, flavour, and crispness. The maximum benefit is achieved by using quantities up to 5·0% of the flour content. Egg up to 5·0% of the flour can be used to advantage, as it helps to strengthen the wafer sheet and reduces the tendency to stick to the wafer plates. The advantage of egg is partially due to its emulsifying properties which can be simulated by substituting the egg with lecithin. Lecithin should be dispersed in oil before addition and is added at the rate of 2 oz for every 1 lb egg (1 : 8) it replaces. The use of melted fat or oil is optional, but increases the tenderness of the wafer sheet and reduces the tendency to stick. Quantities over 2·5% of the flour are of little value. A maximum sugar addition of 4·0% of the flour will increase the crispness of the sheet, but will hasten discoloration from caramelisation.

Aeration of the wafer sheet is caused mainly by the sudden production of steam when the hot plates are in contact with the thin layer of batter. It is necessary to use sodium bicarbonate to correct the acid nature of the flour and milk powder. If the batter is acid, it does not flow or spread to fill the plates properly. The best results are achieved if the batter is pH 7·0 (neutral). If it becomes alkaline, the sheet readily discolours, and the appearance and flavour deteriorate. Although the suggested quantity of sodium bicarbonate is 0·2 to 0·3%, it may be necessary to increase, depending upon acidity and strength of the flour, and on the baking speed. Vol may be used to adjust the aeration as and if required.

When wafer sheets are to be sandwiched with cream, there is no point in using any flavouring in the batter, as it can be added to the cream, but wafers intended for use on their own will be improved by the addition of some vanilla flavour (approximately 0·031% vanillin crystals).

Methods of mixing

As the wafer mixing is a batter, either a batter-mixing machine or a low-geared whisking machine is necessary, to achieve a smooth, free-flowing batter, suitable for depositing on and covering, wafer plates. There are basically two methods:
 (a) Blend all the dry ingredients well together in the machine and while still mixing, run in the liquid ingredients. Continue blending until the mixing is of the desired consistency and is thoroughly blended.
 (b) Dissolve all the soluble ingredients in the liquid in the machine and while still mixing, run in the flour. Continue as in method (a).

Baking of wafer sheets will vary, according to the type of wafer oven and method of firing, and to the thickness, area, and weight of the sheet. The baking period and temperature should be the minimum required to completely bake and dry out the sheet, but not to discolour it. After baking, the sheet should be allowed a very short period in which to cool and should then be creamed and sandwiched. Cooling, sawing to size, and packing, should follow as soon as possible. Owing to the low moisture content of the wafer sheet and its hygroscopic nature it quickly absorbs moisture from the atmosphere. As the moisture content increases, the wafer sheet distorts or buckles, causing sandwiched wafers to split and unsandwiched sheets to be wasted. If the moisture content is permitted to exceed 2·0%, the wafer becomes tough and quite unpalatable, therefore, once the wafer is baked, the time lag before it is packed in moistureproof containers, must be extremely short.

Cream fillings for wafers are usually based on the standard recipe of 60 parts icing sugar to 40 parts fat, such as hardened coconut oil or hardened palm kernel oil. Colour and flavours are added as required. Creams will be considered later in this chapter.

MARSHMALLOW

Marshmallow is a mechanically aerated foam, composed of sugars in solution and including a foaming and stabilising agent, usually of a protein nature. Originally, marshmallow was produced by boiling sugar in solution and adding it to a foam made from albumen and stabilised by the use of agar-agar. This process has subsequently been simplified and the ingredient cost has been reduced by omitting the boiling process and the albumen. The agar-agar is normally replaced by gelatine. The following formulae are typical examples of both varieties:

MARSHMALLOW (original)

Sugar	45·0
Glucose	7·4
Albumen	15·0
Water	32·0
Agar-agar	0·6
	Total 100·0

Method: The agar-agar is soaked in approximately two-thirds of the water for at least 20 hr. Dissolve the sugar and glucose in the remaining water and boil to a temperature of 115°C (240°F). Add to this syrup the strained agar-agar solution and boil to 107°C (225°F). While this solution is boiling, the albumen is whisked to a stiffish meringue, the boiling syrup poured steadily into the meringue, and whisking continues until the mallow is stiff and cool. This is a basic or stock marshmallow. For use, to every 100 parts stock mallow add 31·25 parts by weight of albumen and whisk to a light foam. This type of marshmallow should be light and tender when set.

MARSHMALLOW (standard)

Sugar	30·0
Glucose	25·0
Invert syrup	23·0
Gelatine	2·0
Water	20·0
Total	100·0

Whether manufacturing marshmallow in small batches or in bulk, the mixing process consists of preparing a syrup of the sugars and some of the water, and dissolving the gelatine in the remaining water. These are then brought together and whisking commences. (When preparing a bulk syrup it may be more convenient to dissolve the gelatine in the water and then add the sugars. Only one vessel is required for preparing the syrup in this manner.) The syrup temperature should be prepared at a temperature of 37-43°C (100-110°F), and maintained at this temperature until required. To avoid fermentation all equipment must be sterile, the sugar content should be in the region of 70·0% and the syrup should not be made up in too large quantities, so that it does not stand for long periods. Newly made syrups should not be mixed with syrups made previously.

Colours and essences may be added at the commencement of whisking, but if possible, it is more satisfactory to add towards the end of the whisking time. However, this is only feasible when whisking in open vertical mixers. Water soluble colours and essences are the most satisfactory, and indeed, essential oils must be used with the greatest care, as the introduction of oil or grease in any form is liable to break down the foam structure. Top quality flavours are ideal for use in marshmallow, as the syrup undergoes no heat treatment that will spoil them, and the mallow being slightly acid and a moist medium, they show up to their best advantage.

The methods of mixing the marshmallow syrup vary according to the type of machine in use, but all rely on the incorporation of air bubbles which are finely divided mechanically and are held in position by the protein substance to form a foam or aerosol. In a vertical mixer whisking may take up to 20 min, depending upon the ratio of sugars used, the amount of gelatine, the temperature of the syrup, and the speed and type of mixer. In a pressure whisk, the whisking time may be reduced to approximately 5 min at a pressure of 15 lb psi. Continuous mixers are so efficient that the syrup may pass through the mixing head in a few seconds. One other advantage of continuous mixers, is that the foam produced is controlled at a consistent specific gravity. Whereas with batch production, even if each mixing is at the same specific gravity when mixed, the specific gravity alters during the standing time from start to finish of each batch. This variation naturally leads to inequality in the weight or volume of the product, regardless of adjustments to the depositor.

The balance of ingredients is an important factor in marshmallow production, as the type of mallow can be altered by varying the ingredient ratios. It must be borne in mind, however, that basically, the total moisture content should not exceed 30·0%, and the dry solids content should be at least 70·0%, including approximately 2·0% gelatine. The gelatine content generally depends upon its Bloom strength (see Chapter 7 'Setting materials') and the climatic conditions; as the prevailing temperatures increase, it is frequently necessary to increase the gelatine content to ensure a firm set. The type of sugars used influence the nature of the mallow; sucrose produces a light, tender mallow, which tends to dry out; glucose gives body to mallow, and when used in excess, produces a tough, rubbery mallow; invert syrup, because of its hygroscopic nature, attracts moisture from the atmosphere and keeps open marshmallows moist and palatable. In fully coated mallows, invert syrup in the mallow and the base maintain the moisture equilibrium necessary to prevent splitting. Gelatine, with a high

Bloom strength, or used in excess, will also result in tough, rubbery mallow, irrespective of the glucose content.

Bases of marshmallows, which are to be fully coated with chocolate or other impervious coating, must be correctly matured and conditioned, or the coating will split during storage. The maturing process is purely a matter of allowing the bases to be in contact with the atmosphere to absorb moisture until they have a moisture content of between 5·0 and 6·0%. If the bases have a lower moisture content, they will absorb moisture from the mallow and the biscuit will swell, causing the hard coating to crack or split.

Marshmallow should present a relatively dry surface at the enrober when being fully coated. Moisture used to smooth out the point left by the depositor, must be completely dried away before the coating is applied. If any moisture remains, it may dissolve some of the sugar in the chocolate, and this will crystallise on the surface, causing sugar bloom. Uncoated mallows are usually sprinkled with sugar, coconut, or other shredded nuts, to form a dry exterior to facilitate handling and packing.

Marshmallow is frequently deposited simultaneously with a small quantity of jam or jelly. Although the use of jam or jelly causes greater production problems, it adds to the character and interest, as well as to the flavour of the marshmallow.

JAMS AND JELLIES

Jams and jellies are normally purchased by the biscuit manufacturer, and if obtained from reputable sources, should cause no problems or difficulties. The main requirements of jams and jellies are a good bright colour, a strong pleasing flavour, good setting powers, freedom from seeds, stones, and skins when required for use in a depositor, and a consistent solids/moisture content. For mallow work, the moisture content can be similar to the mallow, i.e. in the region of 30·0%, but for cream and jam biscuits, the moisture content should be no more than 20·0%.

The formulae for jams and jellies vary according to the types of fruit to be used and depending upon their water content and pectin content. Fundamentally, however, to produce 100 parts jam, it is necessary to start with 130 parts sugar and fruit pulp, since 30 parts of water will be lost during boiling. A basic formula would be as follows:

Sugar	65-70
Fruit	40
Water	20-30

The sugar content may contain 10-20 parts glucose. The water content and sugar content depend upon the water content of the fruit pulp. As this increases, the added sugar increases and the added water decreases. Added pectin may be required, according to the type of fruit being used. Apples, gooseberries, and red currants are rich in pectin and should not require supplementing. Most other fruits will require additional pectin. Citric or tartaric acid may also be required, again depending upon the natural acidity of the fruit. The acid is necessary to assist the pectin in setting the jam, and also to invert some of the sugar. This helps to prevent recrystallisation during storage, which may cause grittiness in the jam. The process of boiling the jam varies with different fruits, but usually consists of ensuring that all the sugar is dissolved before boiling commences. Boiling should proceed as rapidly as possible until the temperature reaches 104-107°C (220-225°F). The jam should be tested for its setting properties by cooling a small quantity quickly. If it sets satisfactorily, no further boiling of the bulk should be necessary. When liquid pectin is used, it should be added during the last few minutes of boiling time. Powdered pectin has to be well sieved with a quantity of sugar to ensure that it dissolves completely and evenly.

The correctly boiled jam is allowed to cool a little, before being filled into sterilised containers. The surface of the hot jam is then covered with a waxed paper disc, some of the wax melts, and when cool, sets to form an impervious layer across the jam. This will prevent airborne infection from mould spores. The jam should have a sufficiently high sugar content to prevent fermentation, but mould would thrive, particularly in the humid conditions arising after the lid is in place.

To ensure a good bright colour, the jam should be boiled and cooled as rapidly as possible. It is sometimes helpful to include liquid food colours at the latter stage of boiling.

In the manufacture of jellies, there are basically two varieties. Those made with fruit juice, and imitation ones made with

BISCUIT WAFERS, FILLINGS 155

water, colour, and flavour. The latter are inferior in flavour, but are cheaper to make. Both types may be thickened or set by the use of pectin or by agar-agar. When using fruit juice, it is usual to use pectin. The jellies are usually prepared from equal weights of fruit juice and sugar, or the sugar may be slightly in excess of the juice (up to 20 parts glucose may replace sugar). The pectin and acid quantities will depend upon the type of fruit in use. In the imitation jellies, pectin has to be added, powdered pectin is most suitable, as the liquid pectin is rather dark in colour and may spoil the clarity required of a jelly. Citric or tartaric acid are both suitable. When using colours, they must be water soluble, with a good, bright, clear colour, and must be stable in the presence of acid and sulphur dioxide (sulphur dioxide may be present in the fruit juice as a preservative.)

PECTIN JELLIES

(a) *Fruit* Sugar   60·0
            Fruit juice   50·0
            Pectin   As required
            Acid   As required
            Colour, if required

(b) *Imitation* Sugar   65-75
                Water   25-30
                Pectin powder   1·0
                Acid   0·5 (dissolve in water 1·0)
                Colour   As required
                Flavour   As required

Method: The powdered pectin is well sieved together with a portion of the sugar, the remaining sugar is dissolved in the liquid and is brought to the boil, the pectin-sugar is added and dissolved. The syrup is boiled to 104°C (220°F), and the dissolved acid, colour, and flavours are added. When thoroughly dispersed, the syrup is run into sterilised containers, which are then covered as in jam making.

AGAR-AGAR JELLIES

(a) *Fruit*
- Sugar — 60·0
- Fruit juice — 50·0
- Agar-agar — 0·75
- Acid — 0·2
- Colour — As required

(b) *Imitation*
- Sugar — 70·0
- Water — 40·0
- Agar-agar — 0·75
- Acid — 0·2
- Colour — As required
- Flavour — As required

Method: The agar-agar must be soaked for a minimum of 12 hr before being brought to the boil and simmered for 3 min, and is then strained through a fine sieve. Dissolve the sugar in the liquid, bring to the boil, continue boiling to 104°C (220°F). Allow to cool a little and add the dissolved acid, colour, and flavours. When thoroughly blended, pour into sterilised containers and cover with a disc of waxed paper and finally a lid.

CREAM FILLINGS

Cream fillings of all types are based upon the standard ratio of 60 parts icing sugar or pulverised icing to 40 parts fat. The fat should have a sharp melting point and be brittle when solid, the usual fats in use for filling creams are hardened palm kernel oil, hardened coconut oil, and specialised vegetable oils, resembling cocoa butter in their properties. The amount of fat used in the cream influences the cost considerably, and so to produce a cheaper cream, the fat content would be reduced. This also reduces the fluidity of the cream and makes stencilling more difficult. By including lecithin in the basic mix, the fluidity of the cream can be increased, therefore the original fluidity can be maintained when the fat content is reduced by the introduction of lecithin to the mix. The quantity used varies, according to the ambient temperatures, but approximately 0·1%

of the fat and sugar total will be sufficient, the optimum quantity will be determined by experimenting. To ensure complete and even distribution, lecithin should be dispersed in a small quantity of the fat before being added to the mixing, some manufacturers add lecithin to the bulk oil before processing. If this is done it will not be necessary to include it in the formula.

To this basic recipe, colours and flavours are added as required. Oil soluble colours are not permitted in the United Kingdom, and there is a tendency for water soluble colours not to emulsify with the fat, resulting in a cream of pale colour, full of specks of strong colour, which will not break down in the mixing machine. This can be avoided by dispersing the colour thoroughly with the lecithin and a small quantity of fat, before it is mixed in with the bulk of the fat. When this has been achieved, the sugar and other ingredients can be added. This difficulty can also be avoided by the use of dry, cornflour-based, 'non-speck' colours. These should be well blended through the sugar before the fat is added. The addition of flavours should cause no difficulties, but should be associated with the colour. They should blend in readily, and in the case of fruit flavours, it is usual to add a small quantity of powdered citric or tartaric acid (approximately 0·05-0·1% of the fat and sugar total).

The flavours of some creams may be improved by the addition of a small quantity of salt (approximately 0·2% of the fat and sugar total). The basic formula would be as follows:

CREAM FILLING (standard)

Sugar, icing	60·0
Fat	40·0
Lecithin (optional)	0·1
Colour	As required
Flavour	As required
Acid, citric (optional)	0·05-0·1
Salt (optional)	0·2

Variations to this standard recipe can readily be achieved to introduce variety, not only of flavour, but also of texture. In addition to altering the colour and flavours and also the fat:sugar ratio, use can be made of further ingredients, such as milk powder (skim or full cream), melted unsweetened chocolate or cocoa powder, coconut or other shredded nuts, nut pastes or praline. These may be added in sufficient quantity to influence texture or flavour, or both. The moisture content of a filling cream must not be greater than that of the biscuit, otherwise it will cause softening of the biscuit, and moisture will also accelerate the rate at which rancidity occurs. Any additional ingredients, therefore, must be of low moisture content.

The methods of mixing creams vary, according to the ingredients being used, but the mixing operation is only designed to achieve a smooth homogeneous mass, free from lumps and streaks. The type of colour influences the procedure most, when liquid colours are used, they should be blended with the fat before the sugar and other dry ingredients are added. When 'non-speck' colours are used, they should be blended with the sugar and other dry ingredients before the fat and essences are added. For consistency of results and handling, the cream should be mixed to a standard temperature and specific gravity. These may vary considerably, depending upon the types of creaming machines, the type of fat and other ingredients, and upon the creaming room temperatures. When creaming upon a machine that extrudes cream, rather than stencils, it is usual to add ingredients such as milk powder, cornflour, or rice flour, to help to give 'body' or firmness to the cream.

Savoury filling creams must be made without sugar, and so that the cost is not completely uneconomic, the sugar content, or part of it at least, must be replaced by some relatively cheap and inert filling ingredient. Finely ground cream (or savoury) crackers may be used, bulky powdered flavours, such as cheese (although cheese is usually quite expensive), defatted soya flour, or rice flour, can be used to reduce the ingredient costs. Spray-dried, high maltose glucose, would seem to be an excellent bulking material, with a price competitive with sugar and, at the same time, a very low sweetening value.

The colour and flavour will depend upon what is required and the strength of flavour used. The salt content will also vary, according to the type of flavour, but will be in the region of 0·2% of the total fat and filler. Pepper will be approximately half the salt content. Additional flavouring ingredients such as autolysed yeast or yeast extract, may be included. Mono-sodium glutamate is also often beneficial to bring out the flavour in savoury fillings. Care must be exercised when using these, as many savoury flavours will already contain one or both of them.

SAVOURY FILLING (standard)

Fat	60·0
Cracker meal	40·0
Colour	As required
Flavour	As required
Salt	As required
Pepper	As required

SAVOURY FILLING (cheese)

Fat	50·0
Cracker meal	30·0
Cheese powder	20·0
Colour	As required
Flavour	As required
Seasoning	As required

Icings

Hard icing for biscuits was originally made by beating icing sugar and albumen until there was sufficient aeration to cause the icing to become light, but firm. This was reduced to the correct consistency for covering the biscuit base by adding more albumen. At this stage, colour and essence could be added as required. Unfortunately, albumen is expensive, and has been replaced by gelatine and water, which requires a longer beating time to achieve the desired amount of aeration.

ICINGS:

	Albumen	*Gelatine*
Sugar icing	80·0	80·0
Albumen	20·0	–
Water	–	19·0
Gelatine	–	1·0
Colour	As required	As required
Flavour	As required	As required

Method (a): When using albumen, the sugar is added, and both are beaten until firm and light. The consistency is corrected by adding albumen (water may be used at this stage, but there will be a greater loss of volume and consequently, less biscuits will be covered). Colour and flavour, if required, are added at a late stage, but must be thoroughly blended. The addition of a little acetic acid will strengthen the albumen and help to reduce the beating time.

Method (b): The gelatine should be dissolved in approximately two-thirds of the water, at a temperature of 60°C (140°F). After standing for 10-15 min, the gelatine solution is beaten with the sugar until light. The consistency is corrected by the addition of water at 60°C (140°F), and colour and flavour are added as required. This icing must retain a temperature of approximately 43°C (110°F) when in use, otherwise the gelatine will start to set and cause the icing to thicken.

Cream icing

A cream icing can be made using a similar formula to that for a filling cream. To ensure maximum fluidity, however, it may be necessary to increase the fat content a little (or reduce the sugar). The cream must be mixed and used at a temperature of 37°C (100°F) so that the cream is fluid while working and will set quickly on cooling. (The running temperature will vary according to the melting point of the fat.) This cream icing sets when cooled, whereas the hard icing is dried out at a temperature of approximately 76-82°C (170-180°F).

CHAPTER 14

Quality control

FOR quality control to be effective, there must be standards laid down which are efficiently supervised and adhered to. This applies to every stage of biscuit making, from the receipt of raw materials, through mixing, machining, baking, processing, packing, and storage. It is the concern of all supervisory staff and production operatives, as well as the department that bears the name quality control. The extent to which quality control is carried out naturally depends upon the policy of the company, but as biscuit manufacturing becomes increasingly mechanised and automatic, the need for effective quality control becomes more and more obvious, and the standards become less flexible.

Although standards must necessarily be exacting to promote and maintain the desired quality, they should not be completely rigid. They should specify limits according to the tolerance permitted by the subject under surveillance. This is particularly true of raw materials.

RAW MATERIALS

In order to smooth the path of production, quality control should start in the laboratory, with routine as well as unsystematic checks of all raw materials. Some tests for raw materials have already been described, but these were suitable for a factory with little scientific apparatus and for operatives without scientific training. All tests described in this chapter are more complex and require a trained operator and more expensive equipment. Laboratory testing is designed with two objects in view. One is to ensure that the raw materials are according to specification. The other to find out whether modifications will be necessary to the formula or the handling methods of products in which the particular raw materials is included.

Flour

Flour being the basic as well as the the most variable raw material used in biscuit making, there have been many tests and much testing equipment devised to ensure regularity and consistency in production. Tests are usually divided into mechanical and chemical tests. The mechanical tests are designed to evaluate the physical properties of the gluten and the general characteristics of the flour. Chemical tests are analytical and check the flour for treatment by the miller.

Flour and dough testing apparatus were designed originally to evaluate flour and dough for breadmaking, and can be divided basically into two groups: those which measure gluten quality, and those which measure fermentative properties and, in particular, gas production. In biscuit making, although fermentation is important, gas production is a minor consideration and consequently, only those machines which evaluate gluten quality will be discussed.

Brabender farinograph: The farinograph is designed to measure the water absorbing properties of flour and its strength. The apparatus consists of a mixing unit, the paddles of which run at constant speeds, and resistance is recorded through the torque of the free swinging motor and a system of levers on a graph (farinogram) and on a dial. To determine the water absorbing properties, water is added until the 500 mark is reached for soft flours, or the 600 mark in the case of strong flour. When this figure is attained and remains steady, the amount of water used is determined. This quantity is run directly into the second dough, which is used to test the gluten strength. The machine is left running with the second dough in the mixing unit until the resistance falls to the 400 or 500 mark (depending upon the flour type), or until the dough becomes sticky.

Fig. 7 illustrates the structure of a farinogram. A denotes the dough consistency and consequently, the water absorption of the flour. B represents dough development or the mixing time for full gluten development. C indicates dough or gluten stability as time from full development to the point where resistance weakens and the curve falls. With a strong flour, C remains constant over a long period, but it will be very short with a soft weak flour. D is the elasticity or toughness of the gluten—the thicker the curve, the tougher the gluten. E represents the degree of weakening over a given period of say 20 min.

Fig. 8 illustrates farinograms of different flours, and the varying characteristics are revealed. The farinogram of Manitoba

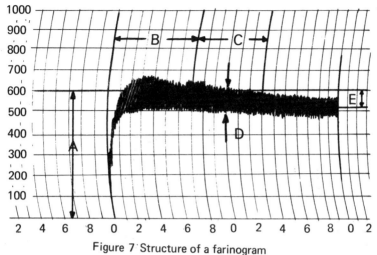

Figure 7 Structure of a farinogram

(a) demonstrates the strength and stability of the gluten, whereas that of strong English (b) shows medium strength without a great deal of stability. That of soft English (c) shows very little strength, very little stability, and considerable weakening.

TYPICAL FARINOGRAMS

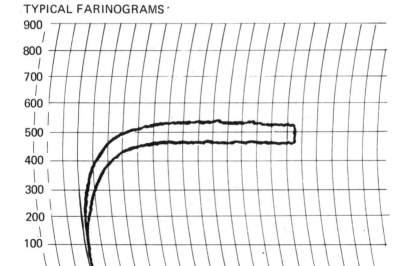

Figure 8 (a) Farinogram: Manitoba flour

164 BISCUIT MANUFACTURE

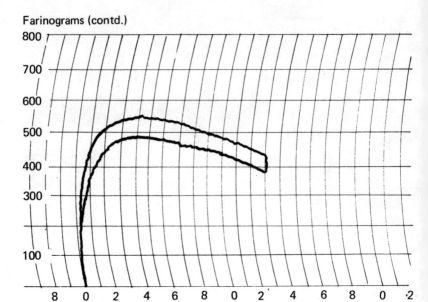

Figure 8 (b). Farinogram: strong English flour

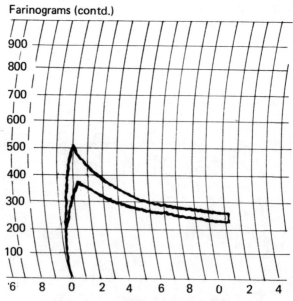

Figure 8 (c). Farinogram: soft English flour

Brabender extensograph: A dough is mixed in the farinograph mixer (mixing time to point B) and is scaled off in 150 g pieces. These are given identical moulding treatment on the extensograph shaper unit. First the dough pieces are moulded into balls, then they are shaped into fingers to fit the cradles of the machine. The cradles containing the dough fingers are placed in the thermostatically controlled proving chambers for a controlled period of time. (The time will vary according to the type of flour under test, but is normally 45 min and multiples of 45 min.) The cradle and dough piece are then placed in position on a swinging, oil-damped, and counter-balanced arm, which is connected to a graph-paper by a recording pen. When the machine is switched on, an arm descending at a constant speed stretches the dough in the cradle until it breaks. The resistance to stretching and the point of breaking are recorded on the extensogram. The dough piece is remoulded and the test is repeated after a relaxation period similar to the initial one has elapsed. The test is repeated once more, giving three curves for the one dough piece, at three different stages in development. From each dough there are three pieces of dough, each of which should give similar curves on the extensogram. The height of the curve denotes resistance or strength, and the length of the curves denotes extensibility. Fig. 9 shows typical extensograms: Manitoba (a) shows high resistance and high extensibility, denoting a strong flour with well-developed gluten. Strong English flour (b) shows only medium resistance but high extensibility. Soft English flour (c) shows very little resistance and high extensibility, denoting that the flour is weak but extensible.

Chopin alveograph (or extensimeter): The alveograph or extensimeter, unlike the farinograph, uses a dough with a predetermined water content, according to a scale and the moisture content of the flour. The total water is usually based on a ratio of flour moisture plus added water to equal 50.0% of the flour weight. Mixing and shaping units are included in the modern alveograph, whereby the dough is mixed and cut into disc-shaped test-pieces under carefully controlled conditions. The discs are allowed to relax in a controlled proving compartment for a given period, according to flour type. The dough discs are then placed one after the other over a fine aperture, through which air passes under pressure. The dough piece is clamped in position with a metal collar, preventing the air from escaping at the sides and therefore causing the dough to stretch in a bubble formation. The air necessary to blow the bubble is recorded on a graph against the time taken before the bubble bursts and allows the air to escape. Thus the alveogram records resistance and extensibility, denoting the strength and suitability of the flour. Fig. 10 shows typical alveograms: Manitoba (a) demonstrates the resistance and stability combined with extensibilty. Strong English (b) shows considerably reduced resistance and stability with medium extensibility. Soft English (c) shows little resistance and stability, but considerable extensibility. (X signifies where the bubble of dough finally collapses.) The curves are taken as an average of each dough piece tested from one dough on the same graph paper.

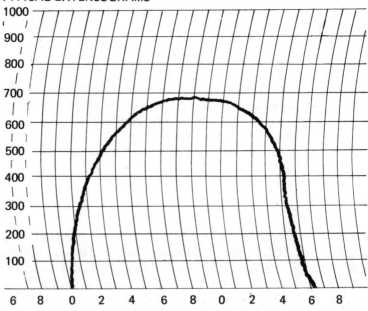

Figure 9 (a). Extensogram: Manitoba flour

Simon 'Research' testing unit: The Simon Research testing unit is also known as the Miller's Research testing unit; it consists of three pieces of equipment, namely: the mixer-shaper unit, the water absorption meter, and the research extensometer.

Extensograms (contd.)

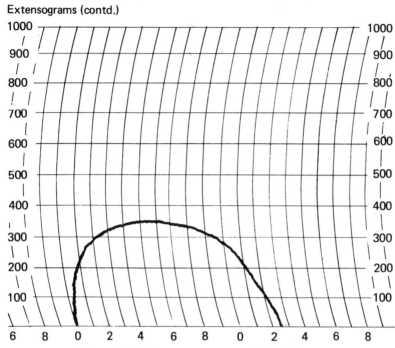

Figure 9 (b). Extensogram: strong English flour

Extensograms (contd.)

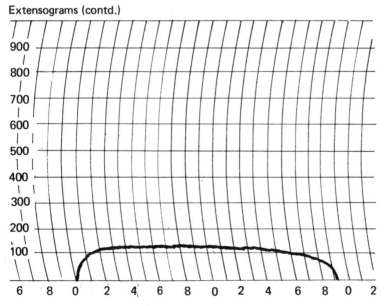

Figure 9 (c). Extensogram: soft English flour

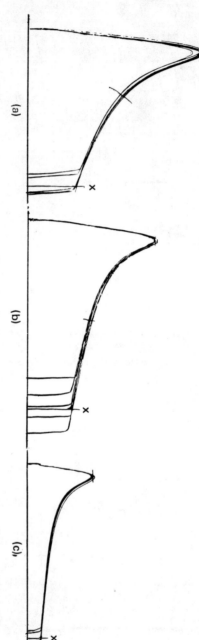

Figure 10. Typical alveograms
(a) Manitoba flour
(b) Strong English flour
(c) Soft English flour

The mixer-shaper unit is designed to provide dough and moulded dough pieces under controlled and standard conditions for testing on the other two pieces of the unit.

The water absorption meter measures the dough consistency of three small doughs mixed in a beaker or, preferably, in the Simon Minorpin mixer. Each dough has a known and slightly different water content. Fermented doughs stand 3 hr, and unfermented doughs stand 1 hr, before being flattened and cut into strips. The strips are rolled lightly and placed in the dough barrel of the meter. The barrel is a small cylinder with an aperture at the base, through which the dough is forced by the pressure of a heavy weight inserted at the upper end of the barrel. A meter measures the rate of dough extrusion from the cylinder, and the time taken for one revolution of the meter is recorded by stopwatch in seconds. The test is repeated with each of the doughs, resulting in different times (the stiffer the dough, the greater the time in seconds), and the results are plotted on special logarithmic graph paper. The graph is joined up and should form a straight line. Where the line crosses the 50 sec mark, the water content is read off and this gives the quantity to be used for that particular flour sample in tests on the extensometer. (The water content is given in gal per sack and as a percentage based on the flour weight.)

Samples of flour are made into doughs on the mixer-shaper unit, according to the results of the tests on the water absorption meter. The mixing and handling techniques vary with unyeasted doughs and fermented doughs, but after moulding into shape on the mixer-shaper unit, and after a period of relaxation, the dough pieces are impaled on a split pin on the extensometer. A motor drives the lower half of the split pin downwards at a constant speed. The upper half of the split pin is connected by levers and an oil damper to a graph where the resistance of the dough is recorded. The test is repeated with each dough piece, to give an average. The dough pieces may also be moulded again and tested after a suitable period of recovery. Fig. 11 illustrates typical graphs recorded by the extensometer, showing resistance by the height, and extensibility by the length. The Manitoba and strong English would follow a 3 hr fermented process, and the soft English a 1 hr unyeasted process. The graph (a) Manitoba shows a combination of high resistance and extensibility. Strong English (b) shows only moderate resistance and extensibility and graph (c) soft

English shows little resistance but high extensibility. It will be noted that there are marked similarities in the graphs recorded by the Brabender extensograph, the Chopin alveograph and the Simon Research unit.

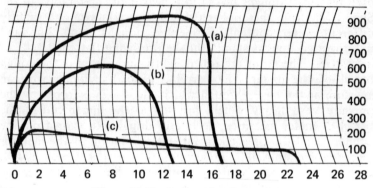

Figure 11. Typical graphs of extensometer
(a) Manitoba flour
(b) Strong English flour
(c) Soft English flour

Gluten estimation

A hand test has been described previously, but there are machines designed to wash away the starch and soluble matter from the gluten. This is achieved by mechanical working of a standard test dough, under a steady stream of water. The dough is paddled in a container with a fine mesh base. The fine mesh prevents pieces of gluten from escaping, but permits the starch and soluble matter to run away to waste. In this way, the process of gluten washing can be standardised. When the gluten sample is obtained it can be assessed for quality and, after drying, the dry gluten content can be calculated.

Many flour specifications quoted by cereal laboratories do not include gluten-forming proteins, but express total proteins as a percentage. This figure includes soluble proteins for which due allowance should be made. The estimation is achieved by determining the nitrogen content of the flour sample, usually by the Kjeldahl method. Nitrogen content is then converted to protein percentage by multiplying by the factor 5·7. (This factor is based upon the normal nitrogen contents of the two main proteins, gliadin and glutenin, present in flour.) The Kjeldahl method consists basically of two stages. First, the flour sample is digested with concentrated sulphuric acid. This, together with other additions, converts the nitrogen into ammonium sulphate. Second, after rendering the mixture alkaline, it is distilled, and the ammonia content is determined by titration of the distillate.

Moisture determination

The moisture content of raw materials is important from both buying and storage viewpoints, but the accurate determination of the moisture content is far from simple. There are various methods, and it is not often that the different methods agree with one another. Therefore, it is very important that a standard procedure is laid down and carefully followed, whichever method is used. Apart from routines already described, there are the following devices and methods:
Distillation.
Drying out - atmospheric pressure, and under vacuum.
Rapid electrical methods: by resistance, dielectric, high-frequency, infra-red heat.
Chemical or titration.

Each of these processes has its peculiar advantages and disadvantages, but any one should be sufficiently accurate for a biscuit factory laboratory, so long as a rigid standard procedure is adhered to.

Ash determination

A more accurate method than that described, using a Bunsen burner, is to ignite the sample in a pre-weighed silica dish, either over a low flame, or in a cool muffle furnace. The muffle furnace temperature should then be increased to, and maintained at 600°C for at least 2 hr. The dish is then cooled in a desiccator and weighed. The ash content is calculated as a percentage of the flour sample. The ash content of pure endosperm is approximately 0·30% and the ash content of pure husk can be as high as 9·0%. It will be seen, therefore, that the more husk or bran is included in the flour, correspondingly the ash content will increase. The inclusion of *craeta preparata* and vitamin mixes will increase the ash content considerably, and should be taken into consideration. Chemical analysis will determine the mineral additives.

Kent-Jones & Martin colour grader: The Pekar method of colour determination is purely a visually comparative one, and depends upon a direct comparison with a known flour sample. The Kent-Jones procedure gives a grade figure standardised against magnesium carbonate, and can therefore be measured and recorded. The figure is obtained by measuring the reflection of a ray of light from a paste of the flour sample, by means of photoelectric cells and a galvanometer. The instrument is designed to measure flour grade, i.e. bran content, and ignores colour diferences as a result of bleaching.

DETECTION OF FLOUR TREATMENT

Bleachers

(a) *Nitrogen peroxide:* To detect nigrogen peroxide bleaching, Griess Illosvay reagent is used. The simplest method is to add one drop of the reagent to each flour sample during a Pekar test, prior to immersion in water. If nitrogen peroxide has been used, a pink colour will result. Alternatively to 10 mls of water and 6 drops reagent, 2 g flour are added, and well shaken in a stoppered test-tube. Again, a pink colour will result if nitrogen peroxide has been used to bleach the flour. The reagent is prepared by dissolving 0·5 g sulphanilic acid in 150 mls acetic acid (30%), and 0·1 g Cleve's acid is dissolved in 20 mls boiling water. The two solutions are mixed together and stored in a well-stoppered bottle. The addition of a little zinc dust, and heating, will remove any pink colouration which may develop. The test should be carried out in a nitrous-fume-free atmosphere (no gas flames). For dependable results, the pink colouration should appear immediately during the test.

(b) *Benzoyl peroxide:* 5 g flour are mixed to a paste in 8 mls potassium iodide solution, and heated on a water bath. The appearance of dark spots indicates the use of benzoyl peroxide (½ oz per 280 lb flour). Alternatively, a wet Pekar slab may be flooded with 0·2% orthotolidine solution. Again, dark spots indicate a positive reaction. If the flour is aged, the peroxide will have decomposed, in which case a more complex test is necessary.

(c) *Chlorine and chlorine dioxide:* The detection of chlorine treatment in flour is not simple, owing to the small quantities involved and the fact that untreated flour has a natural chloride reaction. Chlorine treatment, however, affects the wheat oil which, in untreated flour, is free from chlorine. An ether extract should be made from the flour sample, and the ether evaporated, leaving the wheat oil. A rolled-up piece of copper gauze is heated until it is completely oxidised. This is then dipped in the wheat oil and returned to a low Bunsen flame. If the fat contains chlorine there will be a distinct green flash. This will be more easily observed if the test is performed in the dark.

Mineral stabilisers

Mineral additives are not easy to detect owing to the natural presence of similar salts in flour. It is frequently a matter of comparing the difference between an unknown sample of flour with a known and untreated flour sample. A simple test, known as the chloroform test, will determine the presence of added mineral salts, particularly phosphates and persulphates.

Chloroform test

Ten grammes flour are shaken up with 30 mls chloroform, or carbon tetrachloride, in a boiling tube or separating funnel. On standing, the flour will rise owing to the high specific gravity of the solvent. Tests can then be carried out to determine the nature of the additives.

(a) *Phosphates (which occur naturally):* (1) To a water extract (filtrate) of flour, add a few drops dilute nitric acid and 3 or 4 mls ammonium molybdate solution, and boil. A canary yellow precipitate indicates the presence of phosphates. (Ignore a slightly positive reaction.)

(2) Mix 2 g flour with 20 mls water; add 7 mls strong hydrochloric acid, and 10 mls ammonium molybdate. Add a few drops fresh stannous chloride to 2 mls Fehlings solution, and pour into the flour suspension. A dark blue colour denotes a positive result, while a pale blue colour should be disregarded.

(b) *Sulphates:* To a flour extract, add barium chloride. An intense white precipitate indicates added sulphates.

(c) *Chlorides (which occur naturally):* Add silver nitrate solution to a flour extract. A dense white precipitate indicates a positive result.

(d) *Persulphates:* Prepare a 50/50 mixture of industrial methylated spirits and distilled water (total 100 mls). Dissolve in this 0·25 g p-phenylendiamine hydrochloride. Pour about 5 mls of the solution on to a flattened flour sample. Blue-green specks appearing immediately indicate a positive result. (Use the solution before it is 5 days old.)

(e) *Bromates (potassium):* Pour a freshly mixed solution of 5·0% hydrochloric acid and 1·0% aqueous solution of potassium iodide in equal quantities over a wet Pekar slab. Spots almost black in colour indicate bromate treatment. (Iodates react similarly but give blacker spots.) Dark spots appearing, after a time, indicate added persulphates.

(f) *Ascorbic acid:* Pour a 0·05% solution of dichlorophenolindophenol over a wet Pekar slab and leave for 5 min. White spots, appearing on the blue stained flour, indicate the presence of ascorbic acid.

(g) *Bran:* Pour a 1·0% solution of pyrocatechol in methylated spirits over a wet Pekar slab. This treatment stains the bran, making it more obvious on the surface of the Pekar slab.

Sprouted wheat

If sprouted wheat is included in flour for fermented doughs, the doughs will become sticky and unmachineable owing to the high activity of proteolytic enzymes in the sprouted grain attacking the gluten and rendering it soluble. To detect its presence, prepare a 10·0% flour extract. Place 50 mls of the filtrate in a weighed evaporating dish and evaporate to dryness over a water bath. The dish is then further dried in a drying oven until the weight is constant, and can be expressed as a percentage of soluble extract. This should not exceed 5·0%. If it does, this is a fair indication of sprouted wheats being used in the miller's grist.

Fungal enzymes

The detection of fungal enzyme additives in flour is an extremely delicate process. Briefly it depends upon the destruction of cereal enzymes before tests are conducted for enzymic activity. The cereal enzymes can be destroyed by the addition of certain buffer solutions, to which cereal enzymes are sensitive, and which are harmless to their fungal counterparts. Tests generally need to be conducted for both types on the same sample, in which case the fungal enzymes, if present, should be destroyed, and the cereal enzyme activity can then be compared. Fungal enzymes are more sensitive to heat than are cereal enzymes, and it is possible to inactivate the fungal enzymes while the cereal enzymes remain active. (Bacterial enzymes are normally less sensitive to heat than the cereal enzymes.)

Acidity

Place 5 g flour sample into a 200 mls beaker, and make into a paste with a little CO_2-free distilled water (boil and cool). Add more water to a total of 50 mls and mix well. Add 4 drops 0·5% phenolphthalein indicator and titrate with N/20 sodium hydroxide in a burette. When, in spite of stirring well all the time, a faint pink colour persists for 30 sec, read off the amount of sodium hydroxide used (the titre). The flour acidity is calculated as potassium di-hydrogen phosphate:

$$\% \text{ Flour acidity} = \frac{\text{Titre} \times 0\cdot0068 \times 100}{\text{Weight of flour sample}}$$

The acidity of milk powder can be determined in a similar manner, but is expressed as lactic acid, using this equation:

$$\% \text{ Milk powder acidity} = \frac{\text{Titre} \times 0{\cdot}0045 \times 100}{\text{Weight of milk powder sample}}$$

FATS AND OILS

Second in importance to flour, fats and oils require routine tests. There is unlikely to be any trouble with materials from reputable sources, but there is always a chance of a wrong delivery, or of incorrect processing in the factory. These can be readily checked by tests for the acid value, the Kreis test, the slip melting point, and the specific gravity of the fat before and after processing.

Acid value

The acid value of a fat or oil is a measurement of the amount of free fatty acid present. A low acid value indicates purity of a refined oil, and freshness. If the acid value increases during storage over a period, this is an indication of deterioration. The value is expressed as the number of milligrammes of potassium hydroxide required to neutralise the free fatty acid in 1 g fat or oil. Or it may be expressed as a percentage of the major fatty acid present in the fat.

Five grammes of the fat sample are placed in a clean, dry, conical flask and melted. Add 50 mls of alcohol-ether mixture (composed of 3 parts of 95·5% alcohol to 1 part ether) and a few drops of phenol phthalein solution. The flask is shaken and a sodium hydroxide solution is run in until a faint pink colour persists. The flask should be kept warm and well shaken between each addition of the alkali. The acid value equals the number of mls of N/10 NaOH required to effect neutrality, multiplied by 5·61, and divided by the original weight of fat in grammes.

Kreis test

The Kreis test is a quick colorimetric method of detecting

rancidity in oils and fats. It consists of placing 5 mls oil (or melted fat) and 5 mls hydrochloric acid (specific gravity of 1·19) in a well-stoppered test tube, and shaking vigorously for 30 sec. 5 mls 0·1% phloroglucinol in ether solution are added, and a further 30 sec shaking is given. If a definite pink or red colour develops, then rancidity is present. To determine the degree of rancidity, the fat sample should be diluted at the rate of 1 : 9 and 1 : 19 with liquid paraffin (the fat being the lower ratio). Further tests are continued as above, using the two diluted samples, and the reactions can be compared. If the reaction is positive in the dilutions, rancidity is sufficiently advanced to be dangerous.

Slip melting point

The melting point of a fat is very important, but owing to the different melting points of the component glycerides, it is not easily determined. Briefly however, small quantities of the fat are induced into short lengths of several capillary tubes. The sample must have been kept in a steady temperature for approximately 24 hr. If it is necessary to melt the fat to draw it into the capillary tubes pipette-wise, then the capillary tubes must be kept at an even temperature for at least 24 hr. One of the capillary tubes is attached to a thermometer with the sample of fat close to the bulb. Thermometer and sample are dipped into a beaker or specially designed Thiele tube containing cold water. The top of the capillary tube must be above the level of the water, and the water is then gently heated. The sample of fat must be carefully watched until it becomes soft enough to slide up the capillary tube. At this point the temperature must be recorded, and the temperature at which the fat becomes completely melted is also recorded. The test is repeated with the remaining samples, starting with fresh cold water each time.

Other methods of determining the melting point of fats are used, including the point of solidification and fusion, and the drop method. The points of solidification and fusion are separate tests, the melting part being similar to that already outlined, except that the temperature at which the presence of oil is first noticed is recorded and that at which the sample is completely oil, is also recorded. The point of solidification is determined by stirring a boiling tube of melted fat with a thermometer. The boiling tube stands in a cooling water bath. The temperature will gradually fall and the oil become more viscous and eventually set, whereupon the temperature will rise sharply and the highest point reached at this stage is taken as the solidification point. The drop method consists of placing a sample of fat in a specially designed 'bomb' containing a small aperture. The bomb is suspended in a beaker of cold water which is gradually heated until a drop of oil appears through the aperture. At this point the temperature is noted.

Specific gravity of oils and fats

The specific gravity of fat can be determined in the laboratory, but the most important point of checking this is to keep a constant check on the proportion of air being incorporated during processing of the oils in the biscuit factory. At this stage, it is a far simpler process for the section supervisor to fill a small vessel of known volume and weight with a sample of the processed fat and check for weight. The correct weight of the sample, when the vessel is filled with fat of the correct air content, can be previously determined in the laboratory. Routine laboratory checks should also be made. Precise determinations of the air content can be made by dissolving a weighed sample of fat in warmed paraffin. The released air is measured and expressed as a percentage of the fat.

A test sometimes used, is to test the consistency of the fat. It is the 'Penetrometer' test, in which the depth that an inverted and weighted cone will sink into the sample of fat at a given temperature is measured over a given time. This test can be used to ensure regularity of processing in the factory.

SUGAR AND SUGAR PRODUCTS

Refined sugar has a reputation for purity that leaves little to be desired. Consequently, quality control tests need only be infrequent. The main points to be checked are moisture content, ash content, colour and particle size. These have already been referred to. Raw sugars should be examined for moisture, added colour, and extraneous materials other than sugar.

QUALITY CONTROL

Sugar products are more readily adulterated, and should therefore be more frequently checked—for water content, solids other than sugars, and also for the balance and types of sugars. Jams and jellies should be examined similarly, as well as for pectin, gelatine, and agar-agar. Syrups of all types should be tested regularly with a pocket refractometer, which gives a direct reading of soluble solids present and, by subtraction, the moisture content.

OTHER RAW MATERIALS

Most other raw materials should be tested for water content, and, where necessary, fat content. Materials containing fat should also be tested for rancidity. Raw materials should be checked for freshness of flavour and appearance. Flavouring and colouring materials should be tested for strength in test-mixes, both when received from the manufacturer, and when in use in the factory. All raw materials should be tested for adulteration with inert or filler substances. Chocolate of all grades should be tested for cocoa content, sugar content, and fat content. Chocolate said to contain cocoa butter should be examined to ensure that it does, and the usual routine test of chocolate coatings is for viscosity, since fluctuations can be very expensive. There are various ways of measuring viscosity—from weighing how much flows through an aperture at a constant temperature over a standard time, to how long a controlled amount (volume) takes in seconds to pass through an aperture at a constant temperature. Standard methods of preparation and procedure must be followed. Other methods make use of instruments such as the Brabender viscograph. Although originally intended for measuring the viscosity of cereal product pastes and enzymic activity under increasing temperature conditions, it is well suited to measuring the viscosity of various fluids. The temperature can be set at any level between $20°C$ and $90°C$, or it can be made to increase or decrease at the rate of $1·5°C$ per min. Obviously, for chocolate, a constant temperature would be used. The machine measures the resistance of the fluid in a rotating bowl against immersed blades, and the torque is recorded on a graph. Other viscosimeters are available which are quite suitable for chocolate and similar products.

When the raw materials have been tested and any variations assessed, action can be taken to prevent or minimise variations in the finished product. Wherever possible, the action taken should be restricted to rebalancing existing ingredients without altering the formula itself, or to altering the production process at one stage or another to accommodate the ingredient variation. All alterations should be kept to a minimum. They should not affect the cost or quality, and the standard procedure should be reverted to at the earliest opportunity. It is possible, through gradual changes over a long period, to end up with a product which has little resemblance to the original. This can be averted by promptly returning to the standard procedure at the earliest opportunity. Alterations affecting the cost or quality should only be made at a high level of responsibility, and it is essential that all departments concerned are notified.

Biscuit texture meter

Many attempts have been made to achieve a means of comparing biscuit strength to give an objective result which can be recorded for future reference. As a result of work by the British Baking Industries Research Association (now the Flour Milling and Baking Research Association), such an instrument is commercially available from Baker Perkins Ltd, Peterborough. It consists basically of a clamp which holds about a 2-in (5-cm) stack of biscuits at one end of a pivoted beam. Below the poised stack of biscuits is a small circular saw which is electrically driven. When the saw is switched on, the beam is released and the saw cuts into the biscuits. The time taken for a standard cut to be made is automatically recorded in seconds. Table 5 gives examples of readings from the instrument.

The instrument provides a rapid assessment of hardness, yet does not differentiate between Marie and Shortbread (Table 5), which have markedly different characteristics when eaten. Despite this, it is a useful quality control tool for standardising biscuits, and for recording variations in development work.

(Table 5 is reproduced from Dr Peter Wade's *Rheology and Texture of Foodstuffs,* Monograph No. 27, by kind permission of the Society of Chemical Industry.)

TABLE 5: *Examples of texture meter readings obtained on retail samples of various types of biscuits.*

Biscuit type	Texture meter reading in seconds (average of three)
Butter puff	8
Marie	23
Nice	16
Lincoln	18
Shortcake	18
Shortbread	24
Gingernuts	88
Cream cracker	13
Water biscuit	49

PROCESS CONTROL

The routine quality control of production procedure lies in the supervision of standards applicable to the particular phase of production.

Mixing

Standards for mixing procedure should be related to mixing methods, times and temperatures, and water content (where necessary). Supervision of the correct preparation and the accurate measurement of all ingredients is extremely important and should be checked frequently.

Machining

Machining standards refer to machine settings and speeds, distance between rows, across the band, treatment (wash or sugar), weight of items produced, length of running time per bulk unit, and any special information peculiar to the product.

Baking

Standards for baking refer to baking times and temperatures, fuel pressures, particular oven settings and conditions, distance apart or speed of product, weight before and after baking, size, area and volume of product, moisture content when baked, and colour and appearance of the product. Standards should also include details relevant to cooling and stacking, if these are within the ovensman's province.

Processing

Processing standards are similar to those of machining and baking, with their own special requirements replacing those of baking (drying or cooling as the case may be). In addition there are the important standards regarding percentages and weights of fillings and coatings.

Packaging

Standards for packaging should specify the type of pack, including minimum, maximum, and average weights and sizes, packing materials (quality and quantity), arrangements of pack, if assorted (including detailed replica), outer package and labelling requirements with date code, and any relevant or particular additional information.

Storage

Whenever possible, storage standards should indicate ideal temperature and humidity conditions for the stockroom. Details regarding date codes and maximum storage periods, stock rotation and handling methods, should be laid down. Periodic spot checks of a variety of lines in different packs, and from different dates of production, should be inspected by either quality control or by a panel drawn up specifically for this purpose. The packs should be inspected and reported upon, according to the various standards applicable, and also regarding the general appearance, condition, and taste. These checks not only narrow the incidence of despatch of poor quality products, but also ensure the careful observance of strick stock rotation. The drawing up of standards requires a considerable amount of work based on experience, records, and requirements, but the work will be wasted if the standards are not followed, and it is the duty of section supervisors, quality control, and management, to see that they are. The standards required must be impressed upon the operatives concerned, and they should be encouraged to draw the attention of authority when the standards are not fulfilled. The sooner sub-standard products are reported, the sooner can faults be rectified and the lower the wastage (or the lower is the despatch to customers of products of poor standard). It is customary for customer complaints to be dealt with by the Sales department, but the offending products should be returned to the quality control section for investigation and report. This action will bring to the attention of all concerned evidence that unsatisfactory products have been sent from the factory for sale to the public, and the need to avoid similar occurrences.

TROUBLE-SHOOTING

Trouble-shooting is the term given by Americans to the detection, determination, and elimination or rectification of faults occurring during or after production. Most of the ordinary faults were discussed when we dealt with raw materials, and the influence raw materials have when in dough form. Owing to the wide variety of biscuit types, ingredients, ratios, production methods, and techniques, faults can only be treated in a general way.

(1) *Spread or flow:* Spread is usually caused through excessive softening of the gluten which forms the structure of the biscuit. The softening may be the result of the action of sugar in solution, or alkali (sodium bicarbonate), or by the gluten being excessively weak or weakened through flour treatment. The most usual causes are through excess sugar, or its dissolving too readily. These faults can be remedied by reducing the sugar content; by increasing the particle size; by reducing water wash applied to the surface; or by decreasing the influence of

humidity in the baking chambers. Excessively weak flour will be detected by quality control tests, and trouble could be avoided by blending with a stronger flour. If the sodium bicarbonate is suspect, it must be reduced. Doughs which are already prepared, and are expected to spread, can be held back to their standard size by the inclusion of extra flour, but this action is only acceptable in emergencies, as it alters the basic formula and cost.

To induce flow to biscuits which are below standard size, the amount of sugar in solution should be increased by the opposite procedure to that already outlined.

(2) *Hollow bottoms:* There appear to be many causes of biscuits having bad or hollow bottoms and there may possibly be more than one reason. Hollowness is caused either through gas or air between the biscuit and the steel oven band, or by the biscuit structure being distorted as a result of uneven physical stresses. The major contributing factor to this fault is toughness of the dough. Toughness leads to distortion of the biscuit, either during machining, or when it enters the oven. Toughness can be caused in many ways, but it is the result of over-development of the gluten. Gluten over-development can be the result of the use of too strong flour; mixing the dough longer than is necessary; too little fat. These faults can be remedied by blending with a weaker flour, shortening mixing times, or an increase can be made in the fat content. If a clear dough will not result with a shorter mixing time, the water should be withheld from the dough until the other ingredients are well blended with the flour. The dough will then mix easily, with less opportunity for gluten development.

Toughness can also be diminished by the inclusion of ingredients with a tenderising effect on the gluten (those that induce spread).Although not curing the fault, an increase in sugar or alkalinity will assist in making it less obvious. Distortion may be the result of uneven balance of baking temperatures or oven conditions, particularly in the initial baking stages. Too much top heat may draw the biscuit, while too much bottom heat may set the edges before the biscuit expands. A similar problem may arise if the oven band is very hot at the point where the biscuit is transferred to it from the machine.

Cutters, moulding rollers, dockers, and other parts of the machining equipment, may be the cause of air being trapped under the biscuit, of biscuit distortion, or of gas not being able to escape from under the biscuit. All these possible sources of trouble should be checked. Undissolved chemicals producing over-abundantly gas in one place under the biscuit can be another cause of this fault.

As there are so many possible causes of this particular fault, the usual procedure is to check the oven, the machine, the method of mixing, and any peculiarities or divergencies from standards laid down in each case. Emergency treatment to doughs already prepared, but suspect, is to increase the fat or the sodium bicarbonate, if the fault cannot be rectified elsewhere.

(3) *Checking:* Checking is the term given to the phenomenon of biscuits breaking without being subject to outside force, or developing a fine crack across the biscuit face, rendering the biscuit fragile and unsuitable for handling. Although checking is not fully understood, it is generally agreed to be due to the migration of moisture from the centre towards the edges of the biscuit after baking.

When a biscuit is baked, the outer edges have a lower moisture content then the centre. As the biscuit cools, the moisture is absorbed by the outer edges in an attempt to establish equilibrium throughout the biscuit. As the moisture content decreases in the centre, the structure shrinks, and as the edges absorb the moisture, some expansion occurs, thus setting up internal stresses. If the biscuit structure is insufficiently elastic, a fracture results. To prevent checking, it is necessary either to prevent the migration of the moisture or to produce a biscuit with a sufficiently elastic structure to withstand internal stresses.

As checking is not apparent until approximately 24 hr after baking, it can be a serious and expensive fault if it develops. It is necessary, therefore, to be especially careful in the preventive measures applied to biscuits that are prone to checking. These are biscuits with a low fat content (between 5 to 20% of the flour content), and usually with a low sugar content, and in particular hard dough biscuits of the semi-sweet type. The incidence of checking with unsweetened hard dough and sweet soft dough biscuits, is much lower, and with puff dough biscuits is practically non-existent.

The preventive measures can be applied to the choice and balance of ingredients, mixing and handling methods and baking, cooling and packing conditions.

(a) *Choice of ingredients:* The flour is the most important ingredient, and a great deal depends upon the gluten quality rather than quantity. The gluten should be extensible and not harsh. Harshness and flour treatment to strengthen the gluten help to promote checking. Treatment to render tough gluten more extensible will naturally reduce the probability of checking (Melloene treatment, etc.) The fat, if between the critical 5·0 and 20·0% of the flour, should be a firm plastic fat that will help to give maximum structure and gluten lubrication. The inclusion of some emulsifying agent would be beneficial (e.g., lecithin, soya, glycerol monostearate), to increase the degree of fat dispersion throughout the dough. Sugar should be of fine particle size to ensure maximum solubility and, therefore, maximum gluten softening. Invert sugar has a beneficial action after baking. Owing to its hygroscopic nature, it accelerates moisture migration whilst the biscuit structure is still warm and elastic, and thus reduces the dangers of stresses. The degree of aeration also plays its part. A biscuit with a close heavy texture is more prone to check than one with a light open texture. Chemicals which increase the pH of the biscuit to neutral or slightly alkaline, reduce the checking potential. Ingredients which induce flow, owing to their gluten softening action, also reduce the likelihood of checking.

(b) *Mixing:* Mixing should be continued until the correct degree of gluten development is achieved. Undermixing will leave the gluten structure weak, and overmixing will result in toughness, or even gluten breakdown. As already discussed, toughness can produce hollow bottoms, and biscuits with this fault are also extremely liable to check, owing to the uneven baking and the extra stresses caused by toughness.

(c) *Handling:* Care should be taken to avoid toughening during handling, or setting up uneven or excessive strains in the dough sheet, particularly when lapping on dough brakes. The dough pieces should be reversed in direction as often as possible.

Resting time should be as long as possible. Returned scrap should be kept to a minimum, and adequate 'ripple' should be permitted between the final gauge roll and the cutter. Embossing and engraving on the cutters should not cut too deeply, as they make weak spots in the biscuit structure.

(d) *Baking:* The biscuit should be thoroughly baked to reduce the moisture content and to ensure fairly even distribution of the moisture while the biscuit is still pliable. Checking results more frequently when biscuits are baked over a very short baking time. Spot checks on moisture content after baking can avoid much subsequent trouble.

(e) *Cooling and packing:* Slow cooling helps to retain the biscuit in an elastic state over a long period, during which the moisture distribution achieves equilibrium with less tension or strain occurring. Packing biscuits while still fairly warm will increase the time of structure elasticity, and help to reduce checking. Where there is danger of checking, forced cooling can prove disastrous.

(4) *Fat bloom:* Fat bloom is the term given to the appearance of tiny white spots over the biscuit surface during storage. It is caused through uneven crystallisation of the fat after cooling. To eliminate this, it is necessary to ensure that the fat is being processed correctly in the factory. It is usually accepted that a blend of fats with a reasonably wide difference in melting points, and which will produce a good plastic fat when correctly processed and handled, give no trouble in the biscuit during storage. However, cooling, storage, and retailing conditions, may influence the result. Under these circumstances, the type of blend being used and the processing method should be discussed with the oil refiner, who may be pleased to alter melting points or the sources of the raw materials. Either of these may have a bearing on the appearance of fat blooms. Fat bloom on biscuits should not be confused with fat bloom already discussed in connection with chocolate.

IMPORTANCE OF RECORDS

It is extremely important that a routine function of quality control should be the keeping of records of all alterations and variations made to formulae and processes, and the reasons for the changes. These records will prove useful for future reference when similar alterations are necessary. It is also important that when alterations are required, the rule should be that only one item is altered at a time. Only in this way can the effect of a particular alteration be observed. If two or three changes are made at the same time, it will not be known which was the effective alteration.

CHAPTER 15

Re-use and disposal of unsatisfactory products

IT is not possible to manufacture biscuits without producing some scrap, or biscuits which do not measure up to the standard demanded. The profitability of a certain line may, however, depend upon minimising the scrap produced and its intelligent re-use. Whenever a production line is started up there will be some scrap produced. The quantity will largely depend upon the skill and ability of the operative to correctly reset the equipment involved. Apart from this initial quantity of scrap, with effective quality control and adequate supervision, subsequently the amount should be negligible. The source and cause of scrap should be investigated, and when located it is better to remove the scrap at the source until the cause can be rectified, rather than allow further processing. Scrap biscuits in dough form are more easily used up than if they are baked before removal. Similarly, broken biscuits are more readily disposed of than if they are permitted to pass through the enrober.

The disposal of scrap biscuits of all types depends upon the reason why they fall short of the required standard. It is obviously a better economic proposition to market them if they are of a saleable standard. If they are of a good standard in all aspects except one which renders them unsuitable for the purpose for which they are intended, they may still be suitable for sale through another outlet. For instance, a biscuit may be the wrong weight or size, rendering it unsuitable for a specific pack, but it may be within the limits for inclusion in an assorted pack. Failing this it could possibly be sold in bulk under a secondary label, or as 'broken' to retailers who specialise in cheaper lines. Such offers are particularly attractive to stall-holders trading in markets. Other biscuits of good quality can also be released to the employees through 'staff sales' at prices close to cost. Biscuits unsuitable for sale should be returned for incorporation in similar mixings. Burned or soiled biscuits should not be used again, but should be disposed of as pig food.

Biscuits returned for re-use should be stored in air-tight containers until required and then be treated as follows:

(1) *Sweet and semi-sweet biscuits:* Sweet biscuits should be sorted into their various types, e.g. plain, coconut, fruit, wholemeal, dark, and ginger. They are then reduced to a meal or dust by grinding. This meal can be added to some doughs in the dry state (such as in digestive and ginger), but if it is to be returned to plain doughs, the dust should be mixed to a dough-like consistency with water. This softens the crumbs and they blend more easily into the dough. The amount of water added to the meal should be approximately the same percentage as in the dough. If more water is used, allowance must be made in the doughing-up water. Too much scrap should not be returned to biscuit doughs as this may result in the production of even more scrap, thus causing a vicious circle. The maximum dry weight of scrap should be 5·0% of the flour weight. The main point to watch when using up biscuits of any type is that the broad groups do not become mixed. Biscuits with definite characteristics of colour, flavour, or special added ingredients (fruit, chocolate nibs, or nuts), should be reserved for returning to similar doughs. There should be no necessity to segregate biscuits of different types but of similar flavours.

(2) *Unsweetened biscuits:* Cream and savoury crackers and puff biscuits should be sorted and ground to a fine meal. Although some manufacturers return them to their similar type doughs, it is not an altogether satisfactory practice. It is preferable to use them up in small quantities in sweet short doughs, and owing to their rather hard crisp nature, the best outlet is in doughs of a rough texture (those containing wholemeal flour or broken grain). When savoury creams and fillings are being produced, obviously these types of biscuits are ideal for them as bulk or filler.

(3) *Wafer:* As there is always scrap produced at the wafer saw, the scrap must be kept to the absolute minimum at all other stages of production. The most satisfactory method of using up wafer scrap is by grinding to a smooth paste. The best type of machine for this purpose and for all other biscuits with cream, chocolate, or other fillings of a soft or moist nature, is the 'melangeur'. This consists of two granite rollers on a revolving granite bed which reduces biscuits to a fine paste, which is then returned to the filling cream. The returned scrap should not exceed 5·0% of the total cream (this scrap already contains a

proportion of cream if it is chiefly from the wafer saw section). If too much scrap is returned to the wafer cream, difficulties arise regarding spreading the cream on the wafer sheet by machine, and there is a tendency for the wafers to split owing to a reduction in the adhesion or tenacity of the filling cream. The main difficulty of returning wafer scrap to the wafer cream appears when coloured wafers are in use, but this can be overcome by returning the scrap to a cream of a suitable colour, even if this means returning it to creams for the ordinary biscuit lines.

(4) *Cream biscuits:* Creamed biscuits scrap can be treated in a similar manner to wafer scrap, keeping the different colours and flavours separate. It may be found more satisfactory to grind biscuits containing jam for use in doughs rather than creams.

(5) *Chocolate coated biscuits:* Scrap biscuits, fully or half-coated with chocolate, can be ground to a paste and used in chocolate coloured and flavoured creams. They may also be used in dark biscuit doughs, preferably of a chocolate flavour. Owing to the high costs of chocolate it is imperative that scrap is kept to a minimum, and it is advantageous if there is a profitable outlet rather than grinding up for re-use.

(6) *Iced biscuits:* Because of their high sugar content, iced biscuits should be returned to doughs with some discretion, otherwise there will be danger of the biscuits spreading or taking on excess colour during baking. It is probably advisable to blend iced biscuit meal with unsweetened or semi-sweet biscuit meal before softening with water and returning to sweet doughs. The practice of finely grinding iced biscuits and returning to the icing is not desirable. It results in a very second-rate finish.

(7) *Marshmallow scrap:* As long as marshmallow is free from other products it should be melted back to syrup, rendered sterile, and can then be used in fresh batches of syrup. However, once it is in contact with jam and biscuit bases, this is no longer possible, and the only satisfactory method of using it up is in biscuit doughs. Owing to the high sugar content of mallow it must be used with caution. Wherever possible, the approximate sugar and syrup content should be estimated and an allowance

made in the original recipe corresponding to that which is added. In this way, fairly high quantities can be used up in biscuits with a naturally high sugar content. If dark bases are used in mallow production, then the mallow must be used up in dark-coloured biscuits. Where various colours and flavours are used in marshmallow, the scrap must be used up in biscuits with a strong masking flavour and preferably dark in colour. The obvious answer to all these requirements lies in ginger biscuits. These have a high sugar and syrup content, a strong masking flavour, and dark colour. Mallow can be used up in small proportions in other doughs, preferably of a dark colour.

Ginger biscuits should not be regarded as the 'dust bin' of the biscuit factory. Although the strong flavour of ginger will mask many other flavours, there is a point when these will become evident and, even if the ginger content is increased, one cannot be sure the other flavours will remain unnoticed. Whenever scrap of different flavours is added, the clean distinctive flavour of ginger becomes blurred, but in the initial stages this deterioration of flavour is unimportant and unremarked.

One other method of scrap re-use is to develop a biscuit based upon a large proportion of biscuit waste. It may not be possible to produce a biscuit that will be a top seller, but even if it is used only as a single type in an assorted pack, it will be fulfilling a most important role in diverting a major part of the biscuit waste from the regular, better selling lines.

CHAPTER 16

Development

THE development of new lines is the lifeblood of the biscuit industry under present marketing conditions. The traditional lines maintain their popularity, but it is by the introduction of new lines that new customers and increased turnover are won. It is an economic necessity that development in some form is pursued in a modern biscuit-making company.

Development should be a carefully planned operation. The conception of a type of biscuit that should be popular several months ahead, or even a year later, depends upon market research and farsighted sales direction. Approximate requirements of a biscuit variety and the pack are submitted to the biscuit production and development departments. The development section work on the idea and develop a biscuit to fill the requirements, which has then to be translated into a form suitable for production on the plant available. After small- and full-scale production runs, the samples are submitted back to the original panel in various packs with the relevant data referring to cost, weight, type of pack, and packaging materials. The selected packs are sent on storage and transporting tests and, when passed as satisfactory, wrappers and packing materials are designed and ordered. This complete cycle may take months, and it is unwise to hurry any stage, or costly snags and difficulties may be overlooked which may make all the difference between success and failure.

Although it is suggested that development is directed from a panel, original types of biscuits will be developed from the ideas of the development section itself, and this is to be encouraged. In this case, the sample will be submitted to the panel prior to production trials, and if considered suitable, the production trials will commence, followed by the procedure already outlined.

For successful development it is necessary to have a suitable operative who has certain abilities and facilities. The operative (male or female) must have patience and initiative, must be methodical, and must have experience and knowledge of the raw materials available and the methods of combining them together. It is an advantage to have some experience of biscuit production and processing, but not to be hidebound by traditional methods and recipes, nor to accept too readily the limitations of the plant available. Frequently an operative with a wide background of flour confectionery is ideal for this type of work, as he will have an extensive knowledge of raw materials and different methods of combining them. He will, however, have to appreciate the difficulties of automatic plant production.

The facilities required as a minimum, are a room or section segregated from, but preferably adjacent to, the production department. Equipment should include: benches; a table mixer (5 to 10-qt model); pastry brake and pieces of cutting web to pass through the brake; scales and gramme-scales; raw materials in suitable containers; measures for liquids, various cutters and other pieces of small equipment; shelves and small containers for storing samples for a considerable period; and facilities for recording and filing all recipes, methods, and results of trials. Where processing is necessary, such as baking, enrobing, etc., it is possible to achieve this in the production department at the same time as biscuits of a similar size and nature are being processed.

It is possible to purchase laboratory scale rotary moulders, cutters, and steel or wire band ovens. These will help to reproduce plant conditions and so facilitate the transition from small batch to production batch, but they are expensive and involve considerable capital expenditure.

The methods used in making biscuits on a small scale should be as near as possible to those normally used in production as far as the equipment available will permit. If the normal mixing procedure is an 'all-in' mix, then this method should be adopted on the small scale. All processes should be timed and the temperatures controlled. A basic method for sheeting the dough should be laid down and always followed. If for some reason this is not possible, then deviations and variations should be restricted to the minimum and recorded. Formulae used in development will be much smaller than those used in production. The most readily interchangeable method is to use grammes in development where production use pounds, i.e. if the basic production flour weight is 560 lb, in development, 560 g would be used. (This method depends to a large extent on the equipment in use.) Other simplified methods can be devised, such as dividing by 10 and weighing in ounces instead of pounds. Whatever system is devised, it should bear a direct relationship to the formulae as it would be used in production. To build up an efficient system, it is imperative that detailed records are kept of all trials, whether they are successful or not. Cards should be devised to record all relevant information which can be filed for future reference, and these should be indexed against tins of samples of the actual biscuits (to guard against the deterioration of the samples, a photograph could be included with the record card). The information required on the record card will vary, but should include some, if not all of the following data:

(1) Name of produce and reference number.
(2) Initial formula and date, with room for subsequent variations and development with dates. Results of each trial and variation.
(3) Final formula, and production formula, if different.
(4) Mixing: method, times, temperatures, and variations.
(5) Handling procedure: cutter used, variations, treatment (e.g. sugar sprinkled).
(6) Baking conditions: time, temperatures, etc.
(7) Biscuit dimensions: length x width, length of 20, weight before and after baking, packet lengths and weights.
(8) Reference to sample: types of pack in sample; record of inspection dates and results.
(9) Quality control reference to ingredients used.
(10) Detail and remarks pertaining to appearance, flavour, bite, etc.
(11) Processing: formulae, weights used, percentages, handling methods and variations. (For filling or coating used, e.g. cream filling, chocolate, icing, marshmallow, or jam.) Relevant quality control references.
(12) Panel evaluation report.
(13) Any other comments, reports, or results.

Naturally the more detailed the record card, the more information will be available for future reference, but the clerical work is greater and, depending upon the scale of development work involved, may entail staff for this purpose.

Apart from the senior panel, directing development, there should be a junior panel, supervising development, particularly with reference to original and spontaneous work emanating from the development section. The junior panel should be composed mainly from the various departments of production, possibly with a representative from the sales department and one from the senior panel. In this way the valuable time of the senior panel is not wasted with the unsuitable lines, and has only to deal with its own requests and with those that have been sponsored by the junior panel. The inclusion of non-production department representatives should help to prevent the production representatives from 'killing' promising lines which may present production problems.

Once a line has been passed by the panels, production trials should be proceeded with. Initially, it is more economical to run a small trial on the plant so that the necessary formula adjustments can be made before the cost of a full mixing is involved. From the satisfactory full-scale trial, samples should be prepared in a variety of weights and sizes of packets and packs. These are submitted to the senior panel, who decide upon the final packs, which are then made up in quantity for storage and transport trials. Storage trials should be carried out, not only in good conditions, but also in adverse, and samples should be inspected periodically and checked for condition. Transport trials are intended to show how a certain pack will stand up to the rigours of loading, travelling, off-loading, and changes of temperatures involved during delivery from factory to customer. Unfortunately, as the trial has to be marked for return to the factory, the handlers concerned are aware that the pack is to be inspected at a high management level and consequently, the trial tends to receive preferential treatment. However, to offset this factor, one can take into account that it has to undergo the double journey.

Having successfully passed all trials and panels, standards are laid down, wrappers designed and ordered and, at a suitable time, production and sales departments are ready to launch a new biscuit on the market.

Original development starts with an idea, and may involve different techniques, or different raw materials. Whatever its beginnings, original development is in the minority compared with the development of existing formulae or copying competitor's lines. Development of existing formulae can result in quite interesting lines, and depends upon altering ingredient ratios to produce a biscuit of different character, flavour, or appearance; or adapting a moulded biscuit to suit the wire-cut or rout-press machine and vice-versa. An example of straightforward development of an existing formula is given in Chapter 12, where a basic formula is given for a wire-cut cookie. This is then followed by a series of variations on the same formula to produce six additional lines to the original cookie. The variations could be extended by the addition of minced peel, or chopped cherries (glacé) or chopped angelica. These would add not only to the flavour, but to the colour and appearance of the biscuit. By increasing the sugar content and reducing the fat content, a harder, crisper biscuit would result. It would tend to flow more than the original, but by varying the particle size of the sugar and probably adjusting the aerating chemicals as well, the tendency to flow could be corrected or minimised.

One other form of development was mentioned in the preceding chapter on using up scrap biscuits. Where a certain type of biscuit waste tends to accumulate, a biscuit can be developed specifically to use up this waste. Owing to the poor colour and flavour that normally results in a biscuit containing a very high proportion of scrap biscuits, it is politic to develop a biscuit with a strong, dominant flavour and also a dark, definite colour. If the particular type of excess waste has a characteristic flavour, this should be borne in mind, so that it can be used to its best advantage and not merely masked by another flavour.

PART IV
Plant and equipment

CHAPTER 17

Raw materials storage and handling

MOST raw materials require cool, dry conditions for storage, and should be used in strict rotation according to age. The stock room and methods of storage should be designed to facilitate the observance of hygiene. The handling methods will naturally, of course, have to suit the type of building, situation of the stock room, and the quantity of raw materials being handled. The smaller establishments will probably be able to cope with the intake and distribution of raw materials by the use of pallets and fork lift devices. These will also assist in the rotation of stocks and in moving goods for cleaning under and behind. Goods should never be stored directly on the floor. In the larger factories, the obvious methods of storing and handling raw materials is in bulk.

Bulk storage of raw materials has several advantages, but has disadvantages, too. Raw materials transported and handled in bulk are generally cheaper than those delivered in unit packs. Less labour is required for off-loading and distribution, there should be no wastage from split packages or material left in the containers or any spilt upon the floor when the packages are being emptied, and it is easier to maintain a high standard of hygiene when materials are stored in bulk containers and are delivered by sealed tubes to the point of use. To offset these advantages, there is a high capital outlay in bulk handling equipment, but no capital is tied up in returnable empties. With a smaller labour force, there is no 'pool' to draw from when short-handed on continuous production processes, and there is no revenue from the sale of non-returnable empties.

DRY INGREDIENTS

Flour

As flour is the major raw material used, it is normally the first to be considered for bulk storage. The flour is delivered by bulk tanker in loads of up to 15 tons each ($\cdot 98$ ton = $1 \cdot 000$ kg or 1 tonne). The tanker is coupled to a storage bin, capable of holding the complete load, and the flour is forced from the tanker to the bin by air pressure. The air pressure is supplied either by the tanker or by land-based blowing units. These latter are advisable to minimise noise, as the blowers fitted to the tankers are extremely noisy. Each bin should hold at least one complete load, and the number of bins required will depend on the flour usage and the different types of flour in use. When flour blends are used, it may be necessary to have blending bins too. Wherever possible, there should be holding bins for new deliveries, so that the flour can be tested prior to being passed to the main storage bins, or if the flour is not up to standard, to be returned to the delivery tanker. The most satisfactory storage bins are tall, vertical silos, tapered to the discharge point at the base. These are normally constructed of steel, but are also made from reinforced concrete, lined with timber. The main requirements are non-grip internal surfaces, achieved by polishing and treatment with special coating resins. Thermal and acoustic insulation are usually achieved by the silos being housed in a section or building of their own. As flour is apt to arch after storage for a short while, it is necessary to 'fluidise' the flour by passing air through a porous tile at the base of the silo. In horizontal silos, this problem is greater and the remedy more expensive. Each silo is fitted with a dust extraction unit which removes excessive flour dust to a recovery unit.

The discharge from the bins is normally a worm or screw conveyor. This system may be employed to convey the flour to its ultimate destination, or the flour may be fed into an intermediate hopper from which it is conveyed pneumatically. The worm conveyor system is less expensive to install and run, but is cumbersome, and requires a good deal of attention to avoid infestation. The pneumatic system is more expensive to install and run, but is flexible, unobtrusive, and requires very little attention hygiene- and maintenance-wise. The system operates on pressure from 3 to 20 lb/in^2 through pipes 3 or 4 in diameter. (Lower pressures through larger pipes, may sometimes be used for special purposes such as low feed rates.) The flour is conveyed to an automatic weigher, which may be situated in a central position, with distribution lines running to the different mixing machines, where final batch hoppers are situated, or there may be individual weighing machines for each mixer or pair of mixers. From the final hopper, the flour is gravity fed directly into the mixing machine.

Where flour blends are required, there is a choice of three systems: (1) Flour is fed from the individual storage bins, it is weighed, and each type lies in layers in the final hopper. This method gives an accurate blend, but the only blending action is when the flour is discharged from the final hopper to the mixing machine and during the mixing process. (2) The flour is treated in a similar manner to the first method but, instead of the unblended flours being fed directly to the mixer, they are fed to a blending bin and then to the mixer. This method ensures an accurate and perfectly mixed blend, but requires extra plant and processes. (3) Blending can be achieved volumetrically by the use of special rotary valves from the storage bins, permitting the flow of each flour type in turn after short intervals. The flours become perfectly blended while being conveyed (particularly if sieving is included during this stage), but the system relies on the different flours having consistent densities and flow characteristics.

The control of bulk handling installations can be relatively simple, e.g. the mixing machine operator selects his flour requirements on a panel adjacent to the machine. It can, however, be very complex where punch card recipes are fed into a central panel and an operator selects the sequences for all the machines in the factory. There are obviously control systems between these two extremes, depending on the degree of automatic bulk handling and the capital available.

Sugar

Similar systems are employed for sugar to those already described for flour, but with certain variations. Sugar will flow naturally without fluidisation, but is more sensitive to temperature and humidity, and has an inclination to crust. Pneumatic conveying of sugar tends to break down the grains, but this depends upon the length, pressure, and number of bends in the pipe; low pressure conveying, when suitable, is less damaging. Normally, granulated sugar is purchased and processed as required. The processing consists of grinding in a sugar mill and then grading for particle size through sieves. The meshes can be arranged to produce icing for creams, and pulverised and castor for doughs. Granulated can be drawn direct from the storage bin or as the overtails from the grading sieve. During grinding, the icing and pulverised sugar become hot, and if stored in large bins are liable to become hard, so it is often necessary to use the sugar soon after milling and while still warm. This may entail slight alterations to either formula or method, particularly with filling creams, but it is not an insuperable problem. It is possible, of course, to cool the sugar immediately after grading and before storing. The sugars are conveyed and weighed in a similar manner to flour. Where all-in methods are being used for mixing, the sugars can be held in the same final hopper as the flour, but when they are introduced to the mixing at different times, either separate hoppers are necessary or the flour cannot be weighed until the sugar has been emptied from the hopper.

Other powdered ingredients: Chemicals and colours in powdered form can be handled by similar systems built on a smaller scale. If they are to be added to the mix dry, it would be necessary to ensure that they are in pulverised form. Otherwise it would be wiser to follow normal procedure and dissolve them in water.

Danger

When handling any powder in contact with air, there is a possibility of an explosion. This is considered to be greater with sugar than with flour, so it is advisable to ventilate installations to the atmosphere through hinged explosion doors. High temperatures and sparks are likely to induce an explosion, so it is necessary to avoid these factors whenever powders are being conveyed by air. The friction of the powder in the conveying pipe also build up high charges of static electricity, and it is essential that all pipes, bins, and hoppers, are thoroughly earthed. Earthing will not only minimise the likelihood of sparks, but will also protect operatives from shocks.

LIQUID INGREDIENTS

Fats and oils

Fats and oils are delivered in bulk liquid form in tanker loads of approximately 13 tons. The tanker pumps the oil into storage

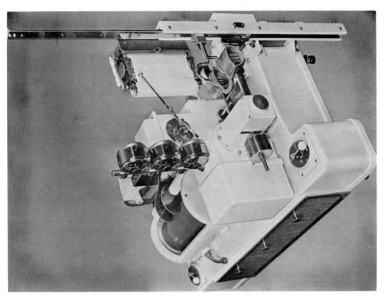

Plate 17b. Brabender Extensograph

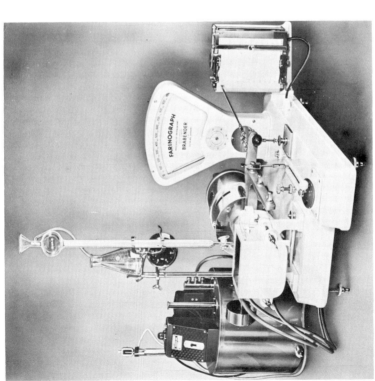

Plate 17a. Brabender Farinograph

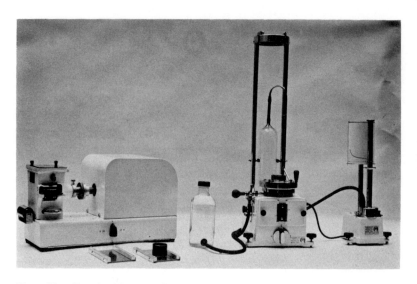

Plate 18a. Chopin Alveograph

Plate 18b. Simon Research Testing Unit

Plate 19a. Baker Perkins Texture Meter

Plate 19b. Baker Perkins Texture Meter showing interior

Plate 20a. Douglas bulk fat storage vessels

Plate 20b. Douglas fat processing equipment with mixing tanks on the right and the emulsifier-cooler on the left

Plate 21a. Douglas silos for storage of plasticised blended fats

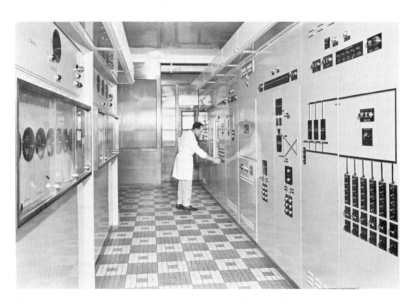

Plate 21b. Control room for bulk handling of ingredients

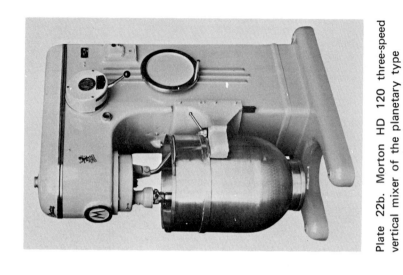

Plate 22b. Morton HD 120 three-speed vertical mixer of the planetary type

Plate 22a. Syrup blending tanks

Plate 23. Simon-Vicars two-spindle vertical mixers

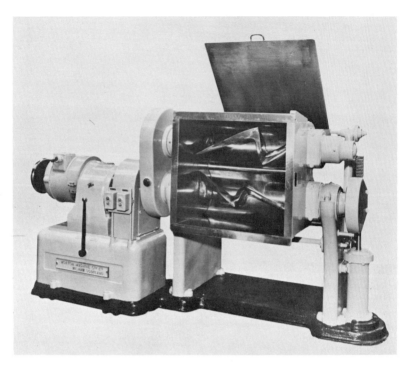

Plate 24a. Morton Duplex 4½ horizontal 'Z' blade mixer

Plate 24b. Morton Gridlap GL 70 horizontal mixer with control console

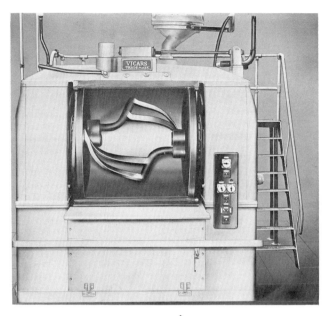

Plate 25a. Simon-Vicars high-speed mixer showing interior

Plate 25b. Simon-Vicars high-speed mixer ejecting biscuit dough directly into a floor mounted hopper

Plate 26a. The Oakes continuous mixer/modifier for bread and biscuit doughs

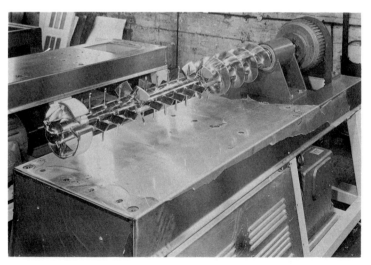

Plate 26b. Mixing rotor shaft of the Oakes continuous mixer/modifier

Plate 27a. The Oakes continuous automatic mixer for batters, marshmallow and fluid mixings

Plate 27b. Stator and rotor of mixing head of the Oakes continuous automatic mixer

Plate 28a. Morton 100 two-speed air pressure whisk

Plate 28b. Simon-Vicars floor mounted tub discharger

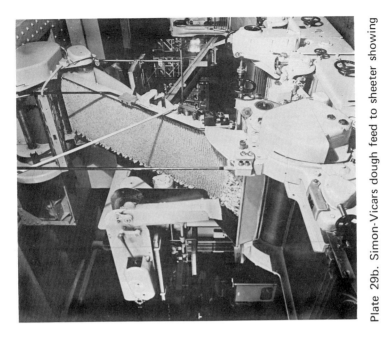

Plate 29b. Simon-Vicars dough feed to sheeter showing scrap return

Plate 29a. Simon-Vicars tub elevator and discharge unit to dough feeder

Plate 30a. Morton heavy duty reversing dough brake

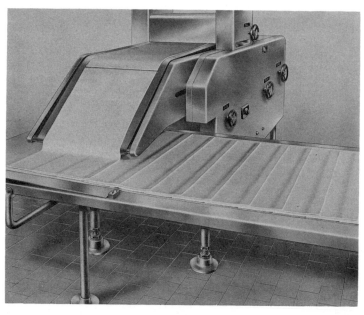

Plate 30b. Simon-Vicars right-angle laminator

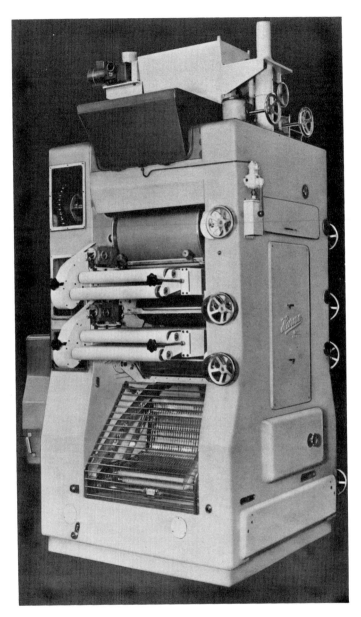

Plate 31. Hecrona vertical laminator

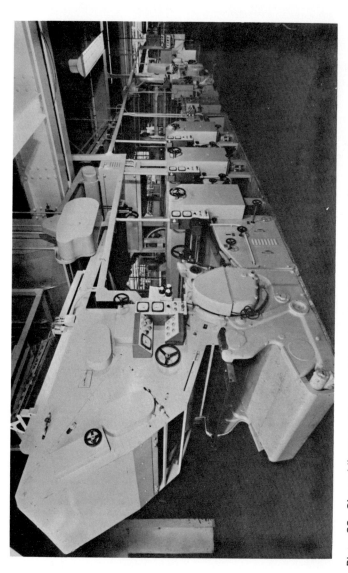

Plate 32. Simon-Vicars cutting machine showing dough feed, pre-sheeter, three pairs of precision gauge rolls and cutting machine cross head

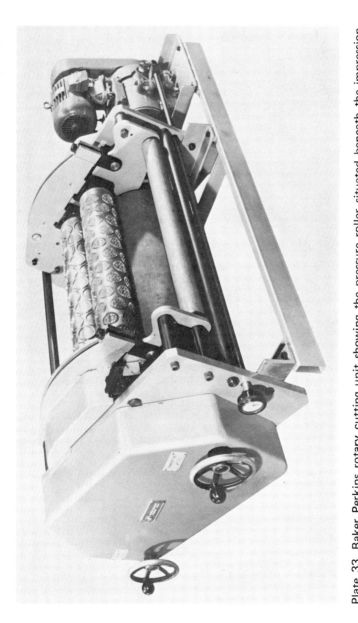

Plate 33. Baker Perkins rotary cutting unit showing the pressure roller situated beneath the impression and cutting rollers

Plate 34a. Simon-Vicars rotary cutting unit in use with the guards removed

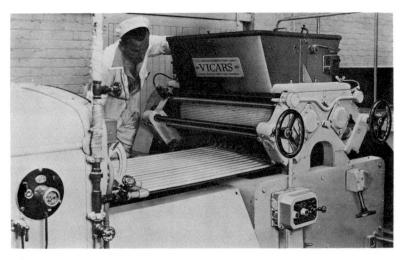

Plate 34b. Simon-Vicars fig bar extrusion unit situated directly above the oven band

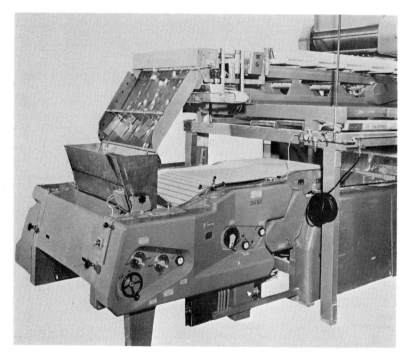

Plate 35. Baker Perkins 88 BT rotary moulder with dough feed

Plate 36. Baker Perkins 88 BT rotary moulder with hopper opened showing the moulding roller and the grooved forcing roller

Plate 37. Baker Perkins Turboradiant oven with direct-fired boost on the first section

Plate 38a. Oil-fired heat sources for one section of a Baker Perkins Turboradiant oven

Plate 38b. Tower-type tensioning for oven band (Simon-Vicars)

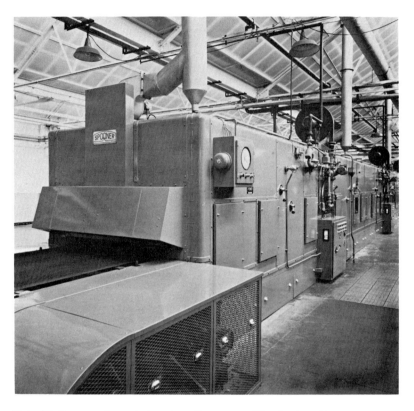
Plate 39. Spooner forced air convection oven with wire band

Plate 40. Oven control panel for a Simon-Vicars gas-fired oven (in the background can be seen a radial reverse turn carrying the biscuits to an overhead cooling conveyor)

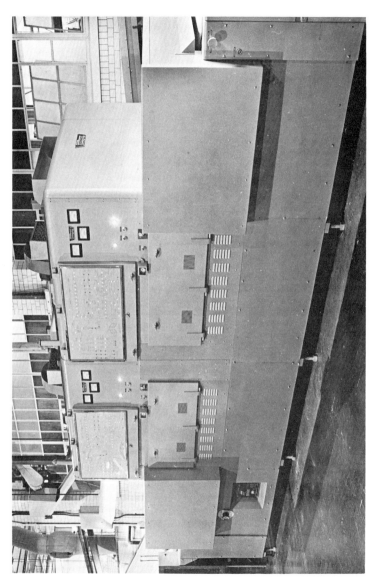

Plate 41. Radyne OCB 301 electronic baking unit

Plate 42. Simon-Vicars Mark II wafer oven

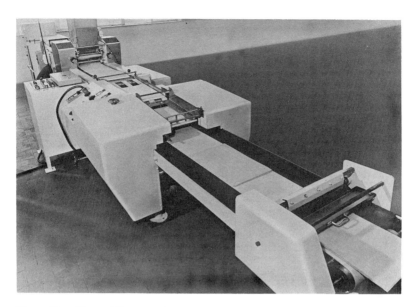

Plate 43a. Simon-Vicars wafer sandwich building machine with creamer in the background

Plate 43b. Simon-Vicars wafer saw

Plate 44. A battery of Simon-Vicars quality creamers being fed directly from the cooling conveyor of the oven

Plate 45. Oakes oven pacer depositor producing marshmallow teacakes

Plate 46a. Walden automatic chocolate kettle

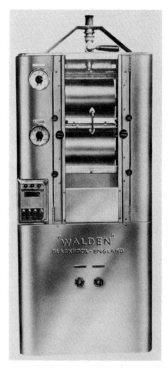

Plate 46b. Walden continuous trickle feed and tempering machine

Plate 47a. Goddard automatic chocolate tempering machine

Plate 47b. The Walden Supreme Mark 7 automatic enrober

Plate 48. Baker-Sollich 76CQ enrober, a sophisticated piece of equipment capable of control by computer

tanks capable of holding a complete load. The storage vessels are constructed of mild steel and are thermostatically controlled with steam or electric heating. Inlet and outlet pipes are usually electrically trace heated. Thick insulation is applied to the tanks to economise on heat, and a temperature between 43 and 54°C (110 and 130°F) is maintained. Sometimes oils and fats are used as they are, but the normal procedure is to process the fat in the factory before use.

Processing is designed to change the hot liquid oil into a cool plastic fat, and is therefore a very important process in biscuit manufacture. It is carried out by pumping the oil or blend of oils from the storage tanks to the blending vessel, where the oils are thoroughly blended and slightly cooled before being pumped to the feeding vessel. The blending and feeding vessels have steam or water jackets for heating or cooling as required, and impellers in the bottom of the vessels to blend and circulate the oil. The oil is pumped from the feeding vessel to the emulsifier cooler. This is a refrigerated barrel through which the oil, and air if used, passes under pressure and is homogenised by rotating beaters, emerging as a fat at a reduced temperature. The times, pressures, and temperatures, vary according to the type of oil, type of chiller, and the use for which the fat is intended. The chiller is designed to produce a fat with a very fine crystal formation, that will remain firm, yet plastic, over a reasonable temperature range and time. The fat may be run into pans at a set weight or may be pumped to a processed fat storage silo. The silo must be maintained at a uniform and carefully selected temperature, depending upon the type of fat in use. The silo is pressurised, and the fat may be metered directly to the mixing machines in water-jacketed pipes or to weighing points. When metering pumps (either time or volume) are used to dispense fat directly to the mxing machines, it is essential that the specific gravity of the fat and the processing and storage temperatures are rigidly maintained, otherwise discrepancies will occur.

Fluid or pumpable shortenings may be purchased, ready for incorporation in doughs, and in this way the processing stage may be dispensed with. The shortenings are stored in bulk, in unpressurised silos, and are then transferred for immediate use by screw conveyors to pressurised silos for distribution to the mixing machines. Blends for these shortenings are similar to those which would be used in the factory, using oils of low, medium, and high melting points, to obtain a long plastic range and fluidity.

When handling oils and fats, it is important that no contact is made with cuprous metals, owing to their oxidative effects on fats. Iron and mild steel are the normal metals used, except when blends contain water or salt (or both), in which instance, stainless steel should be used.

Syrup, glucose, invert syrup, honey and malt extract.

Each of these syrups can be adapted satisfactorily to bulk storage and handling, but frequently only small usage makes it hardly worth while. The syrup can be obtained in bulk by tanker, from which it is transferred to thermostatically controlled storage tanks. The temperature should be as low as possible to suit production techniques, but high enough to permit efficient pumping. Where only small quantities are used, it may be worth while to adapt the system to suit the 5-cwt drums (1 cwt = 50·8 kg), in which the particular syrup is delivered. The syrups are pumped and metered through jacketed pipes to the points of distribution. Volumetric metering is more accurate than metering by time, owing to variations in viscosity of the syrups.

Water

Water can be supplied directly to the mixing machines from the mains, unless timing meters are used, in which case, the variations in mains pressure must be overcome by using a constant level storage tank to maintain a uniform head of pressure. When metering by volume, this is not important.

The water should be available at predetermined temperatures, otherwise tempering facilities must be provided.

Chocolate

Chocolate is available in bulk in liquid form, delivered in insulated tankers and stored in thermostatically controlled mild-steel storage tanks fitted with agitators, to prevent separation during storage. The chocolate is pumped continuously round a jacketed ring-main, from which it can be drawn off at the various holding kettles, where it is ready for tempering or use, as the case may be. The distribution points can be operated automatically without supervision. At each point a meter is placed to record chocolate usage.

In order to avoid congestion around processing machines, such as creamers, the principle of preparing fillings in one place and pumping them to the place of use can be followed satisfactorily with jams and cream fillings. It is necessary to have a holding hopper with thermostatic control and agitators into which the jam or cream is placed, ready to be pumped, directly through stainless-steel pipes to the hopper on the operating machine. The pipe lines should be as short as possible since, in the interests of hygiene, they should be readily cleaned out and preferably sterilised at the end of each production run.

Ingredient preparation

Certain ingredients do not lend themselves to bulk storage and handling, and some are only used infrequently or in very small quantities. These are weighed by hand, accurately, and added to the mixing by the operator at the appropriate stage. It is vital that these ingredients should be weighed or measured with great care, and that some system of checking is observed, not only to supervise the preparation, but also to ensure the inclusion of the ingredient in the mixing. Wherever possible, ingredients should be sieved immediately prior to being mixed, and metal detectors and ferrous metal extractors should be employed at strategic positions. These precautions will not only help to prevent customer complaints, but can also prevent a good deal of damage occurring to the machines which handle the dough or biscuits.

Where bulk handling is in operation, but automatic dispensation is precluded, the ingredients can be pumped by ring-main to discharge points or holding hoppers over platform scales. The mixing-machine tubs are run on to the scales, and the weight of the empty tub is tared. The cumulative totals of the ingredients involved are indicated successively on the dial of the scale, and the requisite weight is discharged from the appropriate main into the mixing tub. When the ingredients are complete, the tub can then be placed directly under the mixing machine, ready for the mixing process. This system is much less expensive than a fully automatic system, and is simple in operation and supervision. The main drawback is that the system entails additional labour and depends upon the operator observing the rules and weighing accurately.

CHAPTER 18

Mixing room equipment

IT is not possible to standardise or lay down precisely the type of equipment one would expect to find in the mixing room of a biscuit-making factory. There are the usual problems of layout, space, production capacity, and also the firm's policy on what is considered to be the function of the mixing room or department. The mixing room is frequently only directly concerned with the production of biscuit doughs. Other mixing procedures for specialist requirements, such as filling creams, marshmallow, and wafer batter, for example, are dealt with in the separate departments as appropriate. None the less, it is convenient to deal with all items of mixing machinery at this stage. All machines designed for mixing have been developed with the main object of blending together a variety of ingredients to form an homogeneous mass. Even so, there exists a remarkable diversity among the various types of mixers which are available, and this diversity is largely a result of the different aims involved during the blending process. Certain machines are designed for the 'creaming process' of mixing, some for the 'all-in-process'; others are designed for stretching the dough to develop the gluten, and there are models to minimise gluten development. Many mixers are intended for stiff, malleable doughs, others for fluid batters; some to incorporate air into the batter. Still further variations are for volume, speed of mixing, and according to whether batch or continuous in mode of operation.

Mixing machines will usually fit into one of the five following categories:

(1) *Vertical mixers.* This classification usually refers to the vertical position of the mixing arm or arms and can be sub-divided thus:
   (i) Planetary,
   (ii) spindle (2- and 3-spindle types).

(2) *Horizontal mixers:* The mixing arm or arms are in a horizontal position. They can be sub-divided:
   (i) 'Z' blade and 'gridlap' types,
   (ii) high speed.

(3) *Reciprocating arm mixer.* An arm or arms moving in a 'to and fro' direction while the mixing bowl or tub revolves.

(4) *Continuous mixer.* As opposed to mixers in the other categories which mix on a batch process, these are designed to mix continuously. They can be sub-divided:
    (i) Barrel type for doughs,
    (ii) rotor and stator head type for batters and fluid mixings.

(5) *Miscellaneous types.* There are a number of mixing machines available which do not fit into the previous classification, and these include:
    (i) Pressure and power whisks,
    (ii) high speed mixers which have been developed for the Chorleywood Bread Process and have subsequently been utilised and/or adapted for high speed production of doughs and batters.

Vertical mixers of the planetary type are not widely used in biscuit production, except where relatively small mixings are being produced, possibly for special purposes. In general, the planetary vertical mixer has a detachable mixing bowl, and detachable mixing arms. The mixing arms are of three basic designs suited to mixing doughs (dough hook); mixing soft, cake-like batters, or 'creaming' (flat beater); and for whipping air into a liquid medium (whisk). The machines are available in varying sizes from 80 qt capacity (approximately 90 litres) down to 20 qt, including 30, 40, and 60 qt in between. Most machines can be fitted with half-size equipment (bowl and arms) by using an adaptor plate for the smaller bowl. Smaller models are also manufactured for laboratory use. Most models are fitted with a gearbox giving 3 or 4 different speeds for mixing. The term 'planetary' is derived from the action of the beating arm revolving upon its own axis in a circular movement around a central pivot, rather like the Earth rotating on its own axis while in a circular orbit around the Sun. This action gives very thorough coverage to the entire bowl volume.

Planetary machines are used mainly where batches of small size are required, and also where some degree of mechanical aeration is necessary. The dough hook is intended to mix dough and develop the gluten content, but can be used for richer biscuit doughs. The beater has limited use for doughs, owing to its wide surface area causing excessive strain on the motor and gears, but it can be used for creaming and mixing soft batters very satisfactorily. When used in creams on medium and/or high speeds, care must be exercised to prevent over-aeration. The whisk attachment is normally constructed of wires, so that air can be incorporated effectively into, and sub-divided in, fluid media such as sponge mixings and marshmallows.

The most widely used vertical mixers in biscuit factories are the twin- and 3-spindle mixers. These are available with detachable mixing tubs (in sizes range from $\frac{2}{3}$ to 3 sacks of flour, giving a range from 300 lb to 1400 lb dough) [137 kg-635 kg]. As these machines have detachable tubs, their flexibility of production is increased. One tub can be carrying prepared dough, one can be on the mixer, and one can be preparing for the mixer. Mixings can be readily conveyed from one part of the site to another, as the tubs are on wheels. Only one size of bowl will fit one machine, it is not possible to fit half sizes.

When the mixing tub is in place under the machine, the spindles are lowered into the mixing position by means of a small (2-3 hp) auxiliary motor. Each spindle has six high tensile steel blades set one above the other in a broken spiral formation. The lowest blade on each spindle fits closely to the bottom of the tub (located by finely set limit switches), and is angled so that it has a lifting action when rotating. The remaining blades are twisted halfway along their length, so that one half of the blade has a lifting action and the other is forcing the dough down. These actions, linked with the action of the interpassing of the blades of adjacent spindles and the friction of the tub surfaces, produce a cutting, churning, and tearing action, ideally suited to soft doughs. While not ideally suited to gluten development as required by cracker and puff doughs, they are quite effective for this purpose. Spindle rotation is generally in the region of 20 or 30 revolutions per minute and consequently, except in the case of rich soft doughs, mixing times can be quite extended (up to 20 min for a fermented dough). Mixing slowly in this way, does help to maintain a low dough temperature, although if, in the case of semi-sweet hard doughs, mixing proceeds for a really extended period, the dough temperature will increase considerably.

The main driving motor varies in power, according to tub capacity and speed of the spindles, but the electric motor for the smallest machine previously mentioned, is 7 hp, increasing in power to 30 hp for the 3-sack, 3-spindle mixer. All moving parts on the mixers are guarded, and unless the main guard over the mixing tub and spindles is securely in position, the driving motor cannot be switched on. All switches and tub locking device controls are situated together at the side of the machine on the main column.

In addition to being able to mix most biscuit doughs satisfactorily on spindle mixers, these mixers are also suitable for filling creams. When mixing creams, some aeration occurs, but this is readily controlled, and there is little change in temperature.

Horizontal mixers

One thing in common which both sub-divisions of this group have, is that the mixing bowls are not detachable. The contents are removed in one of three ways:

(1) The mixing bowl can be made to tilt so that the opened top faces forward and down to permit the contents to flow out, or to be edged out when the mixing blades are carefully rotated.
(2) The forward facing side of the mixing bowl is a sliding door which can be lifted, and again the dough can be edged out.
(3) When dealing with fairly fluid mixings, a trapdoor exit may be fitted in the base of the mixing bowl, and the mixing will flow through this when opened.

While a non-detachable bowl is a disadvantage when the dough or filling is to be moved to another part of the factory, a distinct advantage of these mixers is that the bowl can be fitted with a water jacket for cooling, or for maintaining a suitably low temperature in the dough during the mixing period. If required, hot water may be used to increase the temperature of the contents. In some cases, a refrigerated jacket can be fitted, but this may lead to problems of frosting in the bowl during mixing, and only in extreme cases can it be preferred to the use of chilled water as a coolant.

Horizontal-type mixers are sub-divided into the two-arm type and the high-speed mixer. The first group is further divided, depending upon the general style of the mixing arms. The more usual style is the 'Z' blade configuration. Two arms of this type are set in a 'U'-shaped bowl, with the bottom of the bowl shaped in two arcs to fit the circle described by the movement of each arm. The arms are so designed that, when rotating at different speeds, they force the mixture from one arm to the other—stretching, forcing, and shearing the ingredients rapidly to form a dough. A good deal of heat may be developed from the friction of the action, which justifies the addition of water-jacketing to the tub for temperature control.

The mixing action gives a very thorough action, and can be used for mixing a wide range of goods, including all types of biscuit doughs and even filling creams. Where a creaming action is required, the 'Z' blades are not so efficient as the 'Gridlap' type of mixing arm. These arms are formed rather like horizontal ladders, or grids, at a slight angle across the bowl and, while rotating, they lift to incorporate air into the mixing and sub-divide the air in a creaming action. They are most suited to mixing creams, batters, and soft doughs, but could be used for small rotary doughs. The capacity of both types of mixers may be up to as much as 1200 lb, but the models which handle around 200 lb dough are the most widely used.

The high-speed mixer is mainly used for mixing biscuit doughs of all types, and can handle up to 1600 lb dough at once They have extremely powerful motors which enable bar-type beaters to rotate in the mixing compartment at high speed. There may be 2, 3, or 4 bars, set in such a manner as to counterbalance each other when mixing. These rotate on a single axis and stretch, force, and shear the ingredients to form a dough very rapidly. The bars pass fairly closely to the mixer walls and, in some cases, stationary bars are placed across the mixing compartment to assist in the stretching and shearing action. A mixing time of less than 4 min is fairly standard, and temperature increase is likely to be considerable if not controlled carefully. High-speed mixers are also fitted with 'figure eight'-type mixing arms. These are shaped to give maximum coverage to the mixing tub, and their action is that of lifting the dough and throwing sideways across the tub to be returned by the other half of the figure eight. High-speed mixers are normally fitted with 2-speed control, the fast speed being twice that of slow speed, ratios of 60 and 30 rpm being general, although variations can be obtained. Most mixers include automatic timing (and/or energy input watt/hr meters), and provision is made for direct feed of dry and fluid ingredients, in addition to an opening for manually fed ingredients.

Reciprocating-arm mixers

Reciprocating-arm mixers are widely used for bread-dough mixing, as their action is one of lifting and stretching, so that thorough gluten development takes place. In a single-arm machine, the time required for a dough to be mixed is approximately 20 min. In a twin-arm mixer, the time may be reduced to 15 min. These mixers are available from ¼-sack up to 2½-sack sizes. Normally, the dough bowls are detachable and can be moved around and used for dough storage during fermentation. In general, however, having been developed for bread-dough mixing, they are not entirely satisfactory for mixing requirements in the biscuit factory.

Continuous mixers

Once the dough, batter, filling, topping, or mallow has been mixed, the process is continuous, almost without exception. If the mixing is produced on a batch process, it must be capable of supplying the plant for a period of 20 min, perhaps longer, before the subsequent mixing is ready to follow. During this time, it is inevitable that both physical and chemical changes occur. These can be loss of moisture, gluten relaxation, sugar passing into solution, neutralisation of acid and alkaline ingredients. However slight these changes are, the mixing will no longer be identical at the end of the period with that at the beginning. If these changes occur, it follows that the running qualities on the machines will alter, and these may be significant enough to cause problems on the machines. To eliminate these changes, the mixings should be manufactured on a continuous basis. Continuous mixing itself presents certain problems. The main problem is one of efficient metering of ingredients to the mixing unit, particularly dry ingredients. To overcome this problem requires considerable capital outlay in sophisticated electronic equipment. (This could be an advantage ultimately, as it is one step nearer to completely automatic control of production.) If the continuous mixer is to be used on one variety of mix only, then the problem of metering is minimised, but when change-overs are necessary, further metering problems arise, and these can lead to delays and loss of flexibility in comparison with batch mixers. There are basically two types of continuous mixers available: the barrel type, which is suitable for mixing doughs, and the rotor and stator head type, which is suitable for fluid mixings such as batters and marshmallows.

The barrel-type, continuous dough mixers, consist mainly of a tube (or barrel), through which the dough ingredients are forced by impellers to form a dough of the required characteristics. Provision is made for metering the flour and dry ingredients into the barrel, and as these mix together and pass through the barrel, the fat and liquid ingredients are metered in as required. The barrel is normally water jacketed for temperature control, and the impellers, stators if fitted, and barrel length, are designed to control the nature of the dough. The mixers, according to size, require a 30-35 hp electric motor to drive them, and are capable of producing up to 100 lb dough per min. The Simon-Strahmann mixer is composed of a series of shear discs in which there are a number of holes of diminishing sizes. A shaft, upon which a series of impeller blades are fitted, passes through the centre of the discs. The blades are situated between each shear disc and, when rotating, they force the dough ingredients through the apertures of the shear discs. The dough is sheared by the blades and forced together through the holes, so that it is constantly being cut and reunited throughout its passage along the length of the barrel. The Oakes continuous dough mixer has fixed stators in the wall of the barrel and a central shaft carrying the rotors which, when rotating, force the dough ingredients against the stators with a shearing action which mixes and develops the dough.

The Oakes continuous automatic mixer of the rotor and stator head type is suited to the production of marshmallow and batters such as cake, sponge, and wafer. The mixing head consists of three parts: rear stator, rotor, and front stator. The stators are bolted together, with the rotor mounted on a revolving shaft between the two stators. The mixing head is small, measuring up to 14 in diameter and 2 in depth. The head is surrounded by a water jacket to facilitate temperature control. The internal faces of the stators and both faces of the rotors are provided with concentric rows of blades which intermesh, with ample clearance, when the rotor is in motion. Provision is made for the prepared syrup, or batter, together with a supply of air under pressure, to enter the mixing head through the centre of the rear stator and flow between the blades of the rear stator and rotor to the outer circumference of the stator cavity. The mixture then flows between the blades of the front stator and rotor to the discharge point in the centre of the front stator. The mixing action is very precise, and can be readily controlled by rotor speed, syrup or batter flow speed, and air pressure, to produce the required specific gravity of the mixing. The aerated mix is very stable and has a smooth and uniform texture. As the mixing head is small (the largest has a capacity of less than 3 qt), the specific gravity can be rapidly adjusted, and when changing colour and/or flavour of a similar mixing, very little waste between mixings will occur. The stainless-steel head can be stripped down for inspection and/or cleaning in a very short time. The Baker Perkins Homogeniser or continuous batter mixer, consists of a two-section mixing head, with a rotor and triple impellers in the inlet section, and the main section contains three more impellers, situated between pierced static plates, through which the batter is forced.

Miscellaneous mixers

Power and pressure whisks are available in many sizes, and are designed for a high degree of emulsification, or where a considerable reduction of specific gravity is required. Power whisks are suitable for batch preparation of wafer batters, where efficient mixing of a relatively fluid mixing is necessary without a high air intake. A power whisk consists basically of a cylindrical mixing container with a multiple whisk arrangement mounted horizontally in the lower portion of the cylinder. The whisk is driven by an electric motor. The container can be tilted to discharge its contents, and sometimes provision is made for discharge from the base of the container for liquid mixings. Pressure whisks are a development of power whisks, designed for a high degree of aeration to be achieved in a short time. They are widely used in the production of marshmallow and sponge mixings. The mixing container is a domed cylinder of stainless-steel, suitable for a working pressure of 22 psi

MIXING ROOM EQUIPMENT

($1 \cdot 5466$ kg/cm^2). A multiple whisk is situated in the lower portion of the cylinder. The mixing is placed in the container, whereupon it is sealed. The desired air pressure is attained by means of an integral compressor and retaining cylinder, and whisking commences. Owing to the air pressure, maximum volume is obtained, and whisking time is reduced to a quarter (or less) of the time taken in a vertical planetary type of machine. On completion of mixing, and while the container is still pressurised, a valve can be opened to allow the mixing to be discharged by the force of the air pressure directly to the hopper of the depositing machine. Air pressure whisks are available from 36-200 qt capacity (40-227 litres). Most machines have hoppers into which ingredients are placed while the previous mixing is in the machine. When the mixing is discharged, the prepared ingredients can be admitted to the mixing container direct through a valve, so that the sealed lid does not require removing after each mixing for refilling. A pressurised hopper is also available for adding ingredients which require incorporation after whisking, but before discharge, without releasing the pressure within the machine.

Since the development of mixing machines suitable for the high-speed development of doughs for the Chorleywood Bread Process, there are a number of high-powered, high-speed mixers of various sizes available, which have been adapted for the production of batters and doughs. Most of these machines are capable of producing thoroughly homogeneous mixings in less than 2 min. Energy input controls, or automatic timing devices, are a standard feature of these machines. 'All-in' batch mixings are the general method of production, and owing to the accurate control of the degree of mixing by the machine, careful standardisation of the resultant dough or batter is possible.

DOUGH HANDLING EQUIPMENT

Most manufacturers provide suitable equipment to transfer doughs from the mixing tub (or dough tub in the case of fixed mixing tub machines) to a hopper, from which it is fed to the biscuit-forming machine. Where the mixing room is situated at a higher level than the machine room, a straightforward tub-tipping device almost inverts the tub over a hopper at floor level. The contents of the tub fall into the hopper, from which pre-sheeting rollers extrude the dough in rough sheet form (or in broken-up form if required) by way of a feed web direct to the dough sheeter or hopper of the machine on the floor below. Where mixing is carried out at the same level as machining, it is usual to lift the tub from floor level by means of a hoist. Situated at the top of the hoist, is a tipping device which causes the tub contents to fall into a hopper above the biscuit-forming machine level, so that the pre-sheeting arrangement can feed direct to the machine concerned.

CHAPTER 19

Machine room equipment

FOR the purposes of this chapter it is assumed that the basic function of the machine room is to transform the raw dough into shaped biscuits ready for baking. The equipment used in the machine room is therefore any which may be used in this transformation. It can be classified in three ways:

(1) Dough preparation equipment,
(2) biscuit-piece formation machinery, and
(3) ancillary treatment devices.

Pre-sheeters, to which reference has been made in the preceding chapter, could be included in the first category, as they are designed to produce a constant feed of dough, usually in a roughly formed sheet for further processing in the machine room. To form a dough sheet of high quality, it is necessary to use a dough sheeter. These usually consist of a grooved roller, which forces the dough from the hopper to pass between two smooth rollers, which are set at a pre-determined, but variable, distance from each other. If the pre-sheeter forms a reasonable dough sheet, it is not necessary to use the grooved forcing roller, and the two smooth rolls can then be used as gauge rolls. When scrap dough is being returned after cutting to the sheeter hopper, the forcing roller is essential to form the best possible face on the dough sheet. Sheet formation is extremely important, and every effort to produce a dough sheet of uniformity, both across and along the sheet, should be made in the interests of economy and control of production. Precision setting and control of the gauge rolls is necessary to minimise weight differences across the sheet, but the biggest variation in weight along the sheet results from specific gravity variations in successive doughs. Gauge rolls are normally set so that the dough sheet is diminished in thickness by half at each pair. Speed controls are independent for each pair and also synchronised (including the conveyor webs) throughout the complete unit. High precision gauge rolls have indicators showing the opening at both ends of the rollers in divisions of 0·001 in (0·025 mm), and a tachometer indicates the running speed in revolutions per minute.

Before passing through the gauge rolls, puff, cracker, and semi-sweet hard doughs are usually laminated to build up layers so that their characteristic texture may be developed. The process of lamination can be achieved by hand, or on automatic laminators. Both processes are briefly described in Chapter 11 (Classification of biscuit types and methods of production). A reversing dough brake is necessary for hand production. This consists basically of two horizontal working platforms, divided across the centre by two smooth rollers one above the other. The direction of rotation of the rollers can be controlled to pass the dough from left to right or vice versa and the upper roller can be lowered or heightened as required. Guards are placed to prevent operatives' hands catching between the rollers, and in many cases lifting or lowering the guards, actuates the direction of roller rotation. Reversing dough brakes are available with synchronised webs on the working platforms to facilitate the feed of dough to, and removal from, the rollers. Some machines are programmed for automatic and gradual reduction of the roller gap with each change of direction.

Automatic laminators vary in design, but have in common two dough hoppers, complete with sheeters, producing two sheets of dough separately. Scrap is normally returned to the hopper which forms the lower sheet, so that the biscuit face is not spoilt by its presence. Situated above the lower sheet is a fat and flour mixture sprinkler, which deposits an even layer on the dough sheet before the upper dough sheet meets to form a 'sandwich'. (Semi-sweet hard doughs would not require this addition and normally only one dough hopper is used.) Generally, two sets of gauge rolls ensure complete binding of the sandwich layers, and also reduction of thickness prior to lamination at an angle of 90° to the machine direction. Lamination of the dough may be in a pendulum fashion, whereby the dough feed reciprocates over the conveyor web with a continuous sheet of dough. As the dough sheet moves away from the laminator towards the cutting machine, the dough builds up in diagonal layers with looped edges. The alternative system is closer to the hand method, whereby the dough sheet is cut into pieces to fit across the conveyor web to the cutting machine. The laminations are square instead of diagonal, and the edges are cleanly cut rather than looped. A further variation on these horizontal laminators, which are set at 90° to the production line, is the vertical laminator. This can be placed in direct line with the plant, with a consequent saving in space. The principles of the machine are basically the same as the horizontal machines, but the hoppers, sheeters, sprinkler, and gauge rolls, are mounted vertically or horizontally according to type, but still at 90° to the production line. The pendulum action laminates at the base of the unit prior to the dough sheet passing to the cutting machine. In each case, the dough sheet must pass through two or three pairs of gauge rolls before reaching the cutting machine. Fig. 12 shows a line drawing of the Simon-Vicars vertically integrated dough feeding, gauging, and laminating unit (single hopper type).

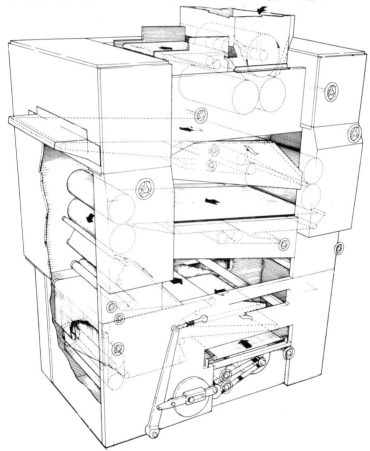

Figure 12. Line drawing of Simon-Vicars vertically integrated dough feeding, gauging and laminating unit.

When puff doughs are to be handled on a laminating machine, sprinkling of fat between the dough layers presents many problems. Consequently, a 'scotch'-type puff dough is made prior to sheeting. In the scotch process, the fat is added to the dough so that the fat remains in fairly coherent pieces. During lapping, layers are built up which are far from continuous, but which are none the less usually adequate to give the required lift during baking. A dough in this state would be extremely difficult to machine, except on manually operated dough brakes. However, if the fat is chilled and minced into small pieces and then thoroughly chilled to ensure hardness of each fat piece, it may be incorporated into a dough during mixing, which could be handled automatically. It is absolutely essential that the dough temperature is low enough to prevent undue softening of the fat, otherwise an homogeneous dough results, preventing satisfactory layering during the laminating process.

BISCUIT PIECE FORMATION

Traditionally, biscuits have been made mainly by cutting the dough sheet with a shaped cutter and also by forcing the dough into an engraved (carved) mould to shape it. The dough piece is then extracted. In both bases some attempt at improving the appearance of the biscuit by embellishment has been made. When biscuits are manufactured in the factory, the same methods are used in a mechanised system.

Cutting machines

Cutting machines are available in two forms:
    (1) Reciprocating or stamping machine, and
    (2) rotary cutter.

The first type is the most widely used, and has been developed to a high degree of mechanical precision. The cutting machine consists basically of three units: the crosshead, incorporating the cutting and embossing crossheads; the cutting table; and the cutter itself. The cutter is attached by bolts to the crosshead mechanism, which causes the cutter to move up and down in a stamping motion. The dough sheet (carried on the cutting web), moves under the cutter and, as the cutter meets the dough and web, it must also move with the dough to prevent arresting the web movement, so that clean-cut biscuits result. The movement of the cutter is in the web direction on the down or cutting stroke, and on the upstroke it returns to the former position. For the cutter to make a satisfactory cut, it must cut against a firm base. This must also be moving in the same direction and at the same speed as the cutter. Consequently, the cutting table is situated directly below the cutter beneath the cutting web. The table not only moves backwards and forwards, synchronised with the cutting speed, but its height is adjustable. It can be lowered so that the cutter does not cut, and be lifted to effect the most efficient use of the cutter. It is usual to cover the cutting table with a pad of thick canvas, or similar material. This helps to reduce wear on the cutting web as it acts as a cushion. Care must be taken to see that the pad is kept in good condition, and that cutter impressions do not become too deep, otherwise the object of the pad will be defeated. Biscuits disfigured through poor quality cutting and too heavy embossing and docking could result. Cutters vary according to their duties, but they generally consist of a series of cutter shells bolted to the cutter frame. Within the cutter shells are brass embossing plates, through which are docking pins. The embossing plates operate independently of the cutting shells, and can be adjusted for length of stroke. Surrounding the cutter shells is a spring-mounted scrap plate, which keeps the shells free of scrap dough. Cutters vary according to their purpose and consequently, cutter design depends upon the type of biscuit for which the cutters are intended. With cracker and puff doughs there is considerable shrinkage after cutting, mainly in the direction of the band, but also across the band. This must be allowed for when the cutter is designed. With square or rectangular biscuits of these types, a 'scrapless' cutter can be made where only the individual divisions across the band and the leading edge of the biscuits are cut at each stroke; the leading edge of the following cut forms the trailing edge of the previous cut. With this type of cutter, the biscuit length can be varied considerably. Only a narrow strip of cuttings or scrap at each edge is returned for reprocessing, hence the name scrapless cutter. The shells for round biscuits from these high shrinkage doughs must be oval (the length of the oval being along the band), so that when the biscuit piece shrinks, it forms a round biscuit. These cutters cannot be scrapless, so it is important that the shells are so placed to achieve the minimum of scrap at each cut.

Shrinkage is minimal with semi-sweet, hard dough biscuits, and little allowance need be made in cutter design. When soft doughs are to be cut, the design considerations may be to leave sufficient room for the biscuit to flow during baking, and greater problems are experienced in lifting soft dough scrap from the web than with hard doughs. It is usual to have 'fingers' at the base of the scrap lifting web to ease the passage of the scrap by maintaining the scrap network intact. These fingers are placed low over the cutting web downstream of the cutter, with one in between each biscuit row and one at each side. Because of these, a double cutter cannot have the cutter shells staggered. Clear lines of demarcation between the rows must be left, otherwise the fingers demolish alternate rows of biscuits.

Cutting machines can operate over a wide range of speeds, but obviously the faster they run, the greater is the wear on the machine and the greater is the noise factor. This can prove very fatiguing to operatives. In many cases double rows of cutter shells may be included in the cutter to reduce high cutting speeds. (This has to depend upon the size of the biscuits and the range of the cutting machines. Further rows can be included with a corresponding decrease in cutter speed.)

Rotary cutting machines have been in use for some time, but owing to the problems of dough piece pick-up and the lack of adjustment in controlling the biscuit appearance, they have not achieved widespread popularity. However, the simplicity of the machinery and the reduction in wear (and noise), leading to lower maintenance costs, have led to the development of a twin roller unit which, it is claimed, eliminates the above problems. This unit consists of two synchronised rollers, which are placed across the cutting web and rotate at the same speed as the web and dough sheet passes beneath. The first roller (impression roller) impresses the biscuit design on the dough sheet (including dockers). This, at the same time, causes the biscuit shape to adhere to the web and corrects the biscuit thickness. The second roller (cutting roller) cuts around the design, the scrap is lifted from the cutting web and the biscuit shapes are then transferred to the oven band, or for further treatment.

MACHINE ROOM EQUIPMENT

Rotary moulders

Rotary moulding machines are suited only to shaping soft dough biscuits. It is usual for the formula to be modified to produce a crumbly dough which will compress into a coherent biscuit shape in the mould, and which will adhere sufficiently to the extraction web so that it is drawn cleanly from the mould for transfer to the oven band. Rotary moulders consist basically of the moulding roller, a grooved forcing roller, and a rubber-clad extraction roller. The moulding roller has a series of biscuit moulds, engraved in the cast bronze face of the roller, into which dough from the hopper is forced by the forcing roller. Excess dough is trimmed from the moulding roller by a steel scraper. Immediately beneath the moulding roller is the adjustable extraction roller around which passes an endless extraction web. The moulding roller and extraction web are in contact under pressure caused by the extraction roller. The biscuit pieces adhere to the extraction web and are withdrawn from the biscuit moulds. Rotary moulders will produce a wide range of soft dough biscuits very efficiently, with a minimum of scrap. Machine running problems are low, and very acceptable biscuits can be produced rapidly and economically. There is some loss in eating quality when a biscuit produced by rotary moulder is compared with a similar biscuit from a cutting machine. This must be attributed mainly to the lower moisture content in the rotary dough, resulting in poorer hydration of the starch content in particular. The main problems arising on the rotary moulding machine are:

(1) The design of the moulds begins to block up and gradually becomes obliterated. This will usually be caused by too dry a dough which lacks coherence under pressure. Other factors which may cause this problem are: too high a temperature of the dough, the machine, or the machine room; too low a slip melting point of the fat. All these could lead to oiling of the dough fat content which will accumulate in the design marking, and prevent easy release of the dough.
(2) Mis-shapen, or incomplete shapes in the mould—usually due to the dough being too soft.
(3) Wedge-shaped biscuits and biscuits with tails, are often associated with dough inconsistencies and also with machine settings. Well-instrumented machines with

variable settings on all the rollers for both speed and pressure, will help to eliminate problems of this nature. The instruments, tachometers, and position indicators, are of great assistance for ease of setting the machine and for standardisation.

Problems may also arise from the behaviour of the extraction web. This web, or apron, is an endless one, usually of fairly open weave, to which the dough pieces adhere after moulding, and during transmission to the oven band. New webs do not always work as efficiently as they might, possibly due to their openness of weave. It is usual to treat the web to improve its performance. Treatment is generally devised to reduce the openness of weave while still retaining its flexibility, and may consist of moistening with water or vegetable oil and rubbing flour into the weave, or even to applying latex to both surfaces. Some machines have built-in thermostatically controlled web conditioners, which incorporate a steam damping device. The extraction web should always be fitted and run with the greatest of care, as a web which is running under uneven tensions, or becomes damaged in any way, rapidly deteriorates and requires replacing. To replace a web at any time is expensive, and if it causes a plant breakdown, it can be exorbitantly expensive. To prevent uneven tensions, it is important that the two rollers (moulding and extraction) are exactly parallel, and that the web is tracked correctly. Excess pressure on the extraction roller, will cause excess wear on the web, and shorten its life significantly. Care must be taken to ensure that the minimum of pressure is exerted to enable efficient extraction of the dough pieces. When the machine is not in use, the pressure of the extraction roller must be released, and the tautness of the web reduced. The scraper, which removes scrap dough from the web on its return journey to the extraction roller, must never be too keenly set, otherwise excess wear will result.

Wire cut and rout press

Wire cut and rout press biscuits are produced on a similar machine, with differing attachments to produce the different types of biscuits. Very interesting and good quality biscuits can be made in these ways but, perhaps owing to the necessary amount of supervision, they are not widely used throughout the industry. Basically, the equipment consists of a hopper containing two grooved forcing rollers which extrude the fairly soft dough through the base of the hopper into which is inserted a die. The die fitting is suitable for the wire cut die, which consists of a series of nozzles, or the rout press die, which consists of a series of cut apertures. In each case the dough is fashioned in a continuous length, according to the shape of the dies, to be cut into individual pieces by one of two methods. The wire cut dies are arranged so that as the dough emerges, a taut wire on a frame (a harp) moves across the face of the dies and severs the dough in a disc. As the wire returns, it is lowered to prevent disturbing the dough which continues to be extruded. The disc of dough falls on to the delivery web and proceeds to the oven band (wire cut machines can be mounted directly over the oven band if required). The wire cut die nozzles may have the centre blocked so that a tube of dough is produced which, when cut, forms a ring. The apertures in rout press dies can be cut in various shapes and sizes to produce many interestingly shaped ribbons of dough when extruded (one shape only per die). The ribbon of dough passes along the delivery web to a cutting or guillotine arrangement, which cuts the ribbon to the required length. The knife edge cuts on the web above a rubber roller, which forms a firm but resilient base to protect the web from damage. The knife is kept free from dough by a pair of close fitting scrapers, through which the knife emerges and withdraws at each stroke. Spacing of the biscuits occurs as they are fed on to the oven band.

Fig bars can be produced by fitting an annular die, through the outer part of which is extruded a tube of dough, and through the centre of which the fig jam is extruded from a separate hopper. Usually, the filled continuous tube is baked, after which it is cut to length by either a guillotine cutter or a circular saw type of mechanism. Baking in a continuous strip gives a more regular biscuit with a moist filling. Shrinkage, or flowing out of the filling, is also eliminated. Very soft doughs can be deposited from the wire cut or rout press type of hopper. Star, or similar shaped nozzles, are fitted to the die and the soft dough is extruded directly on to the oven band or extension. Unless a continuous ribbon of deposited dough is required, the platform on to which the deposit is made must lift up close to the die and then lower so that the extruding dough is broken. In this way a series of star shaped or similar biscuits can be made. A further variation is for the nozzles to rotate or twist as the dough is deposited to yield fancy shapes.

ANCILLARY TREATMENT DEVICES

The main devices used for treatment in the machine room are those which sprinkle sugar and salt or similar ingredients on the biscuit pieces and the wash-over brushes. The sprinkling devices consist of a hopper to contain the ingredient which is to be sprinkled, and, at the base, a grooved roller. When the roller rotates, the granular material is carried from the hopper in the grooves. A spinning brush removes the ingredient from the grooves, and it falls on to an inclined plate which directs the material on to the biscuits which are passing below. The biscuits at this point are carried on a wire conveyor so that excess material falls through the wire and is neither lost nor carried on to the oven band. Beneath the wire is a second hopper to catch the surplus material which, by means of a screw conveyor, is delivered to the operating side of the machine to be caught in a container ready for return to the holding hopper. (Some machines have an integral bucket conveyor which returns the surplus directly to the holding hopper.) To control the amount of material sprinkled by the machine, there is a damper to regulate the degree of contact between the grooved roller and the bulk ingredient. The grooved roller rotates at one of various selected speeds, and the pressure of the brush on the roller can be adjusted as required.

Wash-over brushes hold the liquid wash (water, milk, egg, sugar syrup, or a combination of any of these) in a tank from which the liquid is controlled to maintain a constant level in a shallow tray directly below. A pair of rollers, one picking up the liquid by being partially immersed, transfers the wash to the second roller which is in contact with the rapidly rotating wash brush. This, in turn, transfers the wash to the biscuits. The brush is adjustable for height, which controls the amount of wash transferred. When washes containing milk and/or egg are in use, the utmost attention must be paid to the hygiene of these units. Sterilisation of all parts must be completed daily, as there is a very real risk from the development of food-poisoning organisms.

Flour sprinklers at gauge rolls work in a similar way to that previously outlined, without the recovery system, or by a simpler vibratory sieve mechanism. But these have mainly been superseded by air blowers. Controlled air is directed over the dough sheet just prior to the gauge rolls. This action is sufficient to reduce the moistness of the dough sheet surface and reduces the possibility of the dough sticking to the rolls, yet does not disfigure the dough sheet surface as sprinkled flour may.

CHAPTER 20

Ovens and baking

OVEN capacity is the major controlling factor of production in the biscuit factory. The rate at which biscuits can be baked on a given oven, controls the amount and capacity of the machinery both before and after baking. The oven length and band width, i.e. the baking area in the oven, together with the type of biscuit being baked, control the oven capacity. Ovens are built in varying lengths. Sections of 30 ft (9 m) are fairly standard, and from these ovens can be constructed up to 330 ft (100 m). The oven band can be one of various widths 31½ in (800 mm); 40 in (1000 mm); and 48 in (1200 mm) being the most popular sizes (48 in is now considered to be standard width). The band width, of course, controls the effective width of the rest of the production line. The most usual type of oven band is of cold rolled, hardened and tempered charcoal or carbon steel, which is suitable for a very wide range of biscuit types. Some biscuits, such as crackers of all types, puff biscuits, and some soft-dough biscuits, are traditionally baked on wire bands. These help to retain the biscuit dimensions and allow steam and gas to escape from the bottom of the biscuits as well as the top. When soft dough biscuits baked on a wire band are compared with a similar formula baked on a steel band, the wire band biscuit is markedly denser in texture, but hollow bottoms rarely occur. Wire bands are available in three forms (see Fig. 13):

(1) Open mesh 5 × 5 in. This is in a square weave with 5 wires to the inch in both directions, along and across the band.
(2) Articulated wire mesh. This is not so open as the previous type, and the wire is woven in a zig-zag fashion across the band without any wires running lengthwise along the band.
(3) Wedge wire band, known also as the Continental band. This is closely woven in a diagonal or herringbone pattern, and is suitable for a much wider range of biscuits, including semi-sweet, hard dough biscuits.

Oven bands, steel or wire, pass through the baking chamber and around two large-diameter drums, situated one at each end. The

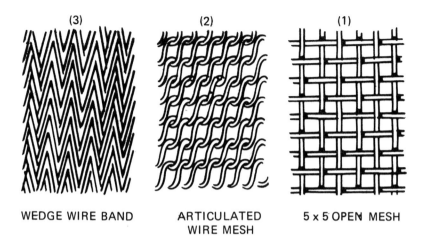

Figure 13. Examples of wire band patterns (Baker Perkins)

band returns beneath the baking chamber, but is usually screened from view by guards or trim sheets. The delivery end drum is counterbalanced to maintain tension on the oven band by either dead-weights or pneumatically. (There is considerable linear expansion when the band is heated from cold to operating temperatures.) The oven band is supported through the baking chamber on skids, and rollers are used to support the returning section of the band. Tracking of the band usually relies upon vertical side rollers, side pulleys, varying the height of the support rollers at one side or the other, or by a combination of these methods. (Side rollers or pulleys cannot be used on wire bands, or the mesh will be quickly damaged). Steel band ovens are normally fitted with a three-brush band cleaning unit, situated beneath the returning oven band. This is followed by a single polisher 'brush'. The 'brush' is composed of strips of material which can be renewed when necessary. Wire band ovens also have the cleaning unit, but have a wire brush mounted inside the band to clean the underside of the mesh and no polishing unit. These units are all independently controlled.

HEAT TRANSFERENCE

To make biscuits palatable, baking is essential, and is achieved by transferring heat from a heat source to the biscuit. Heat can be transferred from one place to another in three ways: by conduction, convection, and radiation. In the baking process, each of these methods plays its part.

Conduction

Heat is conveyed through solid objects by conduction. As each particle of the solid material becomes warm, it vibrates, and being in close contact with its neighbouring particle, heat passes from one to the other. Gradually the whole object becomes warm (or hot) and all the particles are vibrating. The hotter the material becomes, the greater is the amount of vibration, and closely associated with the increase in temperature is the amount of expansion which occurs. Different materials conduct heat at different rates. Copper and aluminium are good conductors (conduct heat rapidly). Wood, asbestos, glass fibre, and air are poor conductors, and are widely used as insulators to prevent heat losses. Insulators are normally porous or fibrous materials, containing a high proportion of air in their structure. The nature of these materials does not assist heat conduction from particle to particle.

Convection

When a liquid or a gas is heated, it expands. As it expands, it becomes less dense. As the liquid or gas closest to the heat sources becomes heated, it expands and becomes less dense and more buoyant than the surrounding material, and begins to rise. It is replaced by the denser material which, in turn, becomes heated and rises. The rising liquid or gas comes in contact with cooler material and consequently cools down (losing heat by conduction), becomes denser, and begins to sink again. Currents, known as convection currents, are set up, and these rise from the heat source, level off and fall, eventually returning to the heat source again to rise, and so on. By this means, liquids and gases become heated.

Radiation

All warm objects emanate heat by means of rays. These rays can travel only in straight lines (unless reflected), and heat objects with which they come in contact. They do not heat the intervening space, and can, in fact, pass through a vacuum, or, in the case of the Sun's rays (the best known example of radiant heat), pass through millions of miles of space to impart their heat upon the Earth. Radiant heat can readily be reflected away if it falls on bright shiny surfaces, but it is readily absorbed by dull rough surfaces.

Heat transference and baking

When an oven is heated, heat is convected from the heat source and the baking chamber becomes hot. Waste gases are carried away through flues also by convection. Convected heat is in close proximity to the biscuits, but its main purposes are to heat the oven chamber or tunnel through which the biscuit passes, and to heat the oven band. Heat is conducted from the oven band through the biscuit pieces from the base, while radiant heat from the fabric of the oven (and from the heat source in direct fired ovens) strikes the biscuit pieces, heating them from above, and penetrates to meet the conducted heat from below. Radiant heat also assists in heating the oven band, and convection through the biscuit piece during baking will depend upon the fluidity of the dough. The three systems of heat transference play their part in the baking process, no matter what the type of fuel or type of oven.

The application of heat to the raw biscuit dough causes many changes to occur. These can be briefly summarised as follows:

(1) *Early stages of baking:* As the temperature of the dough rises, the fat melts, undissolved sugars and chemicals pass into solution, and the whole of the biscuit piece beomes soft and fluid. Gas is produced by the chemicals, and the heat causes it and the air already present, to expand, resulting in the biscuit increasing in volume.

(2) *Mid stages of baking:* As the temperature approaches boiling point, the protein content coagulates and initiates structure formation. This is followed by partial gelatinisation of the starch (there is insufficient water present in the dough for complete gelatinisation). When boiling point is reached, the water content is converted to steam, which assists in increasing the volume, but is mainly lost from the biscuit.

(3) *Final stages of baking:* Biscuit structure is determined at this stage by the coagulated protein, the gelatinised starch, and the very low moisture content. The biscuit is still very delicate and flexible, owing to the fat still being liquid, and the sugars are in syrup form. Surface colour is formed by partial caramelisation of the sugars. (Once the moisture content of the biscuit piece has been driven off at the surfaces, the temperature can rise high enough to permit caramelisation to occur. At the centre of the biscuit the moisture content is normally high enough to prevent discoloration due to sugar caramelisation.)

During cooling, the flexible structure becomes rigid, initially, as the sugar solidifies and later, as the fat solidifies. It is during this period that care must be exercised to ensure that the moisture distribution within the biscuit achieves equilibrium, or equal dispersion throughout the biscuit, otherwise uneven tensions and stresses are set up which may lead to checking. Cooling should be spread over as long a period as possible, and the humidity of the cooling atmosphere should be controlled. Where space does not permit gradual and unforced cooling, filtered air should be blown against the direction of travel, so that it is coolest where the biscuits are coolest, and warmest where the bisuits are still hot, prior to extraction by fan.

The preceding remarks on the changes which occur during baking, refer generally to all biscuits, but obviously where sugar contents are low, the sugar plays a significantly less important role, and the role of the starch content becomes correspondingly more important. In fermented doughs, the structure is entirely dependent upon the gluten-forming proteins and the starch content. During baking of these doughs there is already a proportion of yeast-evolved gas present in the dough, and during the initial stages, the yeast rapidly produces gas until it is destroyed by the heat (temperatures exceeding $52 \cdot 8°C$ ($127°F$), and enzymic activity proceeds until the temperature exceeds $65°C$ ($150°F$)). (See Chapter 12, Classification of biscuit types and methods of production.) Aeration is mainly achieved by the laminating process of building up layers of dough sandwiched with layers of fat and flour mixture. As the dough water is converted to steam it expands, lifting the layers as it does so. The insulated fatty layers prevent a great deal of the steam escaping initially and, as baking continues, the structure becomes set as the protein coagulates and the starch gelatinises. Some aeration is due to the carbon dioxide produced by the yeast, and it is also thought that the gluten content itself bubbles and blisters during baking, before it is coagulated by the heat. The surface colour is normally attributed to the caramelisation of sugars produced during fermentation and baking, particularly as a result of diastatic activity. When milk powder is included in the dough, lactose also contributes to biscuit surface colour, as yeast does not normally ferment lactose.

OVEN HEATING SYSTEMS

The methods of heating biscuit ovens are normally divided into two categories: (1) Direct fired, and (2) Indirect fired.

Direct fired ovens have the heat source located in the baking chamber. For example, a direct fired gas oven would have a series of burners situated above and below the oven band and reaching across the full width of the oven band, spaced at about 2 ft (0·6 m) intervals (see Fig. 14).

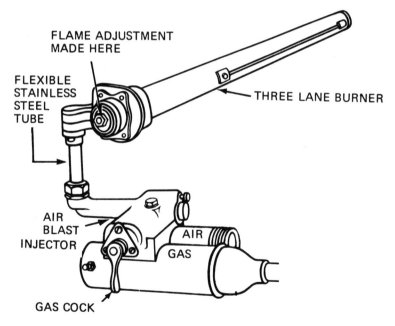

Figure 14. Air blast gas burners for direct fired oven (Baker Perkins)

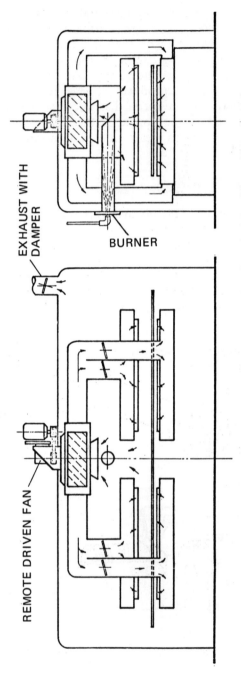

Figure 15. Diagrams illustrating the principle of forced air convection baking used in a Spooner oven section

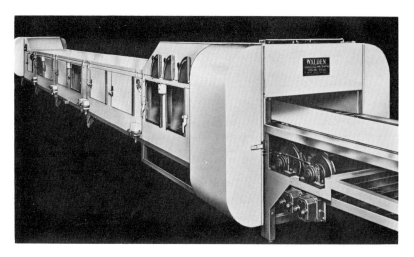

Plate 49a. Walden refrigerated air blast cooling tunnel

Plate 49b. Peck Mix for biscuit crumbing and raw materials sieving

Plate 50. Kek Ltd. laboratory-size grinding mill showing the discs and arrangement of the pegs

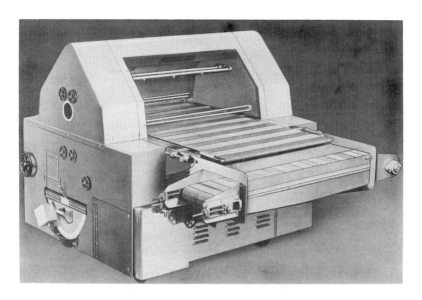

Plate 51a. Simon-Vicars Mark III oil spray unit with cover removed to show interior

Plate 51b. Close up of the Baker Perkins oil spray unit showing the oil dispersers above and below the wire band

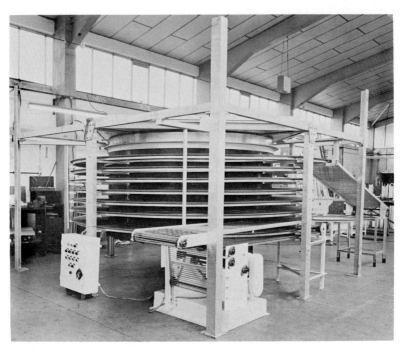

Plate 52. Greer spiral conveying system

Plate 53a. A Lock electronic metal detecting unit monitoring fully coated wafer sandwiches

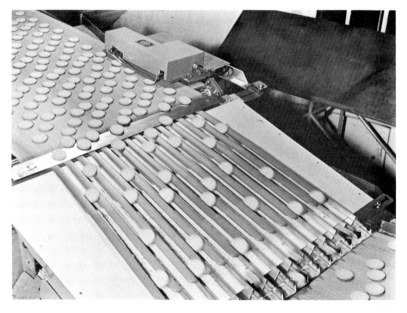

Plate 53b. An example of automatic control by Baker Perkins Developments Ltd. (the instrument in the top centre of the photograph, by means of sensitive feelers, keeps the biscuit rows and the stacker guides in constant alignment)

Plate 54. Aucouturier wrapping machine, fed direct from the cooling conveyor, for round biscuits

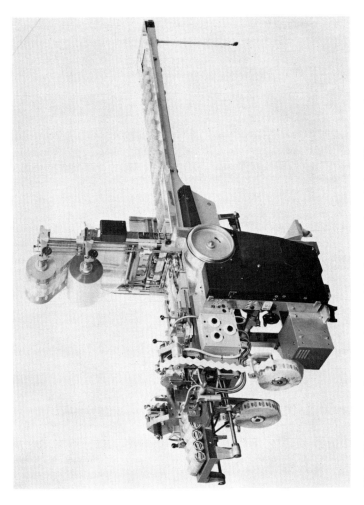

Plate 55. Aucouturier wrapping machine for round packets with end seals

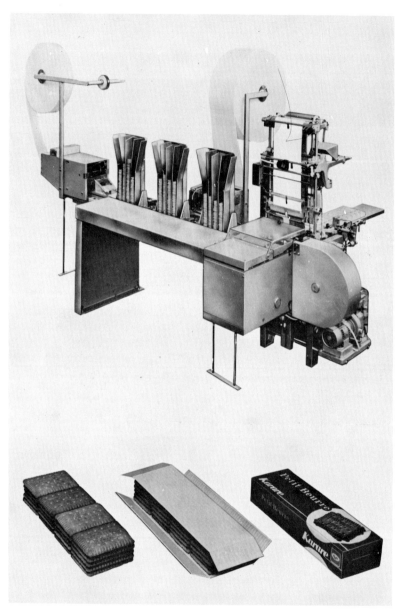

Plate 56. Aucouturier wrapping machine hand fed for square biscuits in piles using corrugated grease-proof paper and a heat-seal overwrap (mode of wrapping is also illustrated)

Plate 57. Rose Forgrove wrapping machine for pillow-type packages

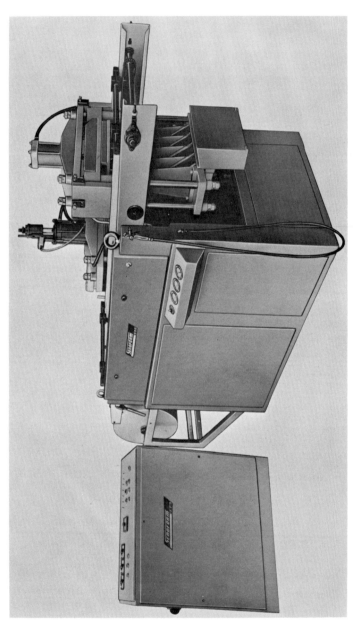

Plate 58. 'Sendform' thermoforming machine for producing moulded plastic packing trays

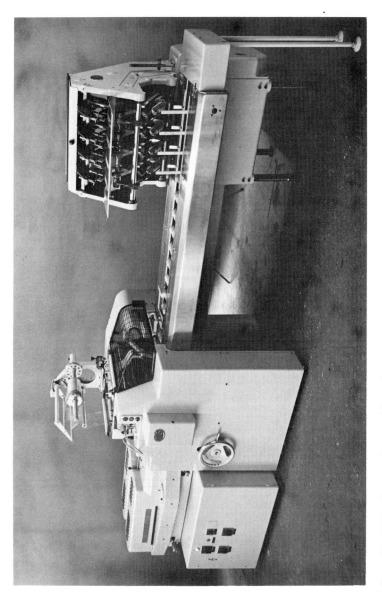

Plate 59. SIG roll wrapping machine with automatic feeder

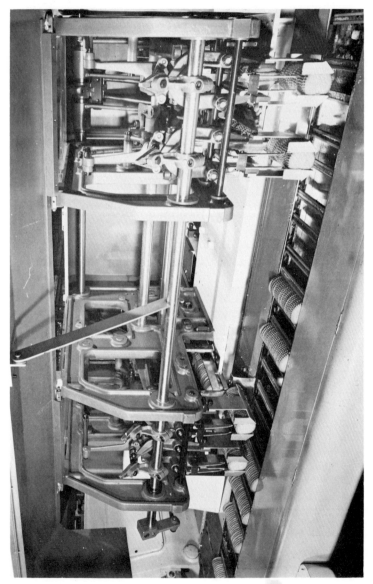

Plate 60. Close up of a SIG automatic loader (type ZH)

Plate 61. SIG automatic wrapping machine for fin-sealed packets with automatic feeder

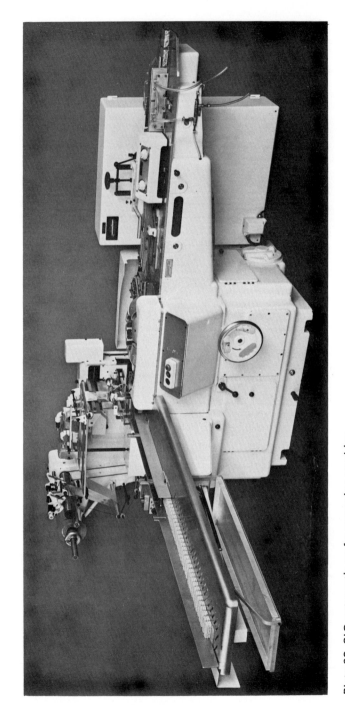

Plate 62. SIG automatic wafer wrapping machine

Plate 63. Jones-Rose constant motion cartoning machine

Plate 64. Rose Forgrove bag forming and filling machine for tumble packs

Indirect fired ovens are heated by means of a high power, single heat source, situated at intervals of approximately 50 ft (15 m). The hot gases produced by the heat source are then circulated by means of powerful fans along ducts through the baking chamber. In some systems, the circulating hot gases are then directed through nozzles from above and below the band to create movement of the heat around the baking biscuits (see Fig. 15). Other systems provide heated air for this purpose, so that the products of combustion are not present at any time in the baking tunnel. The principle of creating movement or turbulence in the baking chamber is that, when biscuits are being baked, water vapour is produced and, as it leaves the biscuit, it forms a barrier between the heat of the oven and the surface of the biscuit. By using forced convection or turbulence in the baking chamber, this water vapour barrier is quickly removed, and the heat can penetrate the biscuit more easily. Reduced baking times are possible in these ovens because of this principle, but alterations in formula may be required to counteract the reduced humidity in the proximity of the biscuit pieces. (Humidity can be controlled by introducing steam into the turbulent air.)

Ovens of both heating systems are available to be heated by any of the following fuels:

Town gas (Coal gas)
Town gas plus methane
Natural gas
Liquid petroleum gases such as butane, propane, and kerosene (paraffin)
Light and heavy fuel oils
Electricity

The type of fuel to be used must be specified prior to the oven being built.

The systems of heating have their advantages and disadvantages, and while manufacturers will supply whichever the customer demands, they tend to have their preferences, and will, of course, stress the advantages of their own preference. For example, while shorter baking times can be achieved with indirect fired ovens of the turbulent and forced convection types, it is usual, for baking cracker-type biscuits, to include direct firing as well in the first section of the oven to boost the heat supplied through the ducts. Similar problems of choice arise when a decision has to be made about the fuel to be used.

The following items should be borne in mind before the decision on fuel is finalised:

Cost of fuel.
Thermal value (heat produced) in relation to cost.
Cost of installation and equipment, e.g. fuel heating, pumps, and compressors.
Cost of storage.
Cost of maintenance and specialist staff.
Availability and reliability of supplies: location of fuel source, location of local depot.
Time involved in attaining required oven temperatures.
Time involved in changing oven temperature at a changeover.
Ease of use: Cleanliness in handling, cleanliness in using, cleanliness of waste products, temperature control during baking.
Safety: Fire hazard, explosion risk, electrocution.
Contamination: of raw materials, of baking goods, from products of combustion.

All ovens using a combustible fuel have built-in safety devices which are designed to prevent fuel being released unless it is correctly supplied with air for combustion and the means of ignition. When the oven is being fired, thermostats guard against overheating (and under-heating). Usually, two or more sections of an oven are combined to form a zone, and an oven normally consists of three zones. Each zone is controlled for heating purposes separately, but by feeding information about temperatures and pressures from each zone to a central control unit, the entire oven can be regulated from one point, or can, in fact, be controlled by computer.

Electronic ovens

It is possible, by means of electronic units, to increase the yield of an oven by up to 40%. These units operate by passing a high frequency electric current through the partly baked biscuits. This causes molecular vibrations within the biscuits and a consequent generation of heat and conversion of water to steam. This system is known as dielectric heating, and is the system of passing an electric current through a non-conducting material, such as a biscuit, to cause heat generation. Normal electricity supplies operate at about 50 cycles per sec, and at this frequency, heat generation is negligible. The electronic units house generators to increase the frequency to as high as 30-40 megacycles per sec (30-40 million cycles per sec), and this is sufficient to achieve the desired heat development in the biscuit.

Electronic units have been tried in each of the three stages (or zones) of baking with varying degrees of success. While the dough is dense and raw, the conventional heating of the biscuit is as successful as high frequency heating. In the second stage of baking, the biscuit structure is extremely delicate, and electronic heating can cause distortion and uneven heating. In the final stage, when the biscuit is porous, and conventional heating systems have difficulty in penetrating, electronic heating can be successful. However, while the biscuits are on the steel band, this has to act as one of the two electrodes, and the high frequency current must be applied in a vertical field. The vertical field has the disadvantage of concentrating on irregularities in the biscuit face, so that biscuits with heavy patterns, rough surfaces, or containing fruit, nut, or chocolate pieces which protrude from the surface, attract the major part of the current to produce uneven baking characteristics. If the current can be applied in a horizontal field these problems do not arise, but the biscuits can not be on a steel band. Consequently, the electronic units are usually placed as near as possible to the delivery end of the oven, immediately after, the biscuits can be stripped from the band. The biscuits are then transferred to a Dacron or Terylene conveyor which passes through the unit, permitting the electrodes to be placed above and below the conveyor in a horizontal field.

The main purpose of these units, placed after the oven, is one of reducing the moisture content to a suitable level. The oven is accelerated to give the desired increase in production, and the biscuits are almost baked. The appearance of the biscuits must be normal, as the electronic unit will not alter their appearance. On emerging from the oven, the moisture content may be 6-7% (or perhaps higher in some cases), but after passing through the short electronic unit (single units may be only 10 ft (3 m) overall length—extra units may be harnessed together as required), the moisture can be reduced to below 2%. Very even moisture contents across and along the band are achieved, as the high frequency current tends to concentrate on high moisture contents. A further advantage of even moisture distribution throughout the biscuit is its reduction of the incidence of checking.

CHAPTER 21

Wafers and second process equipment

WAFER EQUIPMENT

FOR the manufacture of wafers, it is necessary to have a mixing machine capable of producing a batter of high fluidity which has been thoroughly blended to an homogeneous state. Aeration of the batter is unimportant. To achieve the desired batter quality, a power whisk (not a pressure whisk), is adequate, but continuous batter mixers may be used if their high output is warranted by the production capacity of the wafer ovens.

Wafer ovens are available with 12, 18, 24, or 30 plates, and can be heated by gas or electricity. The plates are carried on wheels around an endless track through the insulated main body of the oven. In this area, gas burners are placed to give even heat to both top and base halves of the plates, so that accurately controlled baking occurs. Both halves of the plates are hinged together at one side, in book form, so that they open to receive the charge of wafer batter before baking, and to discharge the baked sheet on emergence from the oven. Wafer ovens have a batter storage tank, from which batter is pumped to a constant level tank. The batter is then accurately delivered by means of a synchronised depositor arm to the base plate. The plates are engraved to give the required design to the wafer sheets, and can be carefully adjusted to control wafer thickness. The depositor partially spreads the batter on the plate, and as the top plate closes on the base plate, the spreading is completed by pressure. The amount of batter requires accurate regulation to prevent waste from incomplete sheets and from excess batter, escaping through the vent holes along the sides of the plates. The vent holes also permit steam to escape from the baking sheet. The production of steam in the sheets is the main means of aeration, and can be quite violent. A steam release device, whereby the top plate lifts slightly in the early stages of baking, allows the initial production of steam to escape readily.

As the plates return to the filling point, the 'book' opens to reveal the wafer sheet and simultaneously, air jets release the sheet which slides down the moving guides of the take-off unit.

The air jets, and the brush under which the sheet passes on the take-off unit, remove any bubbles of scrap from the sides of the wafer sheets. The wafer sheet proceeds then to a collector box which, when full, is used to feed the next machine in the line. Alternatively the sheet may be carried singly in a vertical position for a few seconds only, so that any remaining steam may escape and for the wafer to cool.

Electrically heated wafer ovens have electrical elements built into each plate (top and base), to give even and over-all heating. The current is automatically controlled through a voltage regulator, and the output can be infinitely variable as required, so that accurate and efficient baking is possible. The power is transferred from a series of 'live' bars, mounted in the oven framework to the moving plates by means of carbon brushes. Safety devices cut off the current when the machine is stopped, or access doors are opened.

Wafers which are to be creamed and sandwiched are fed in a continuous line along a plastic-coated web or a steel band beneath a creaming machine. The creaming machine consists of a hopper, at the base of which are two forcing rollers, which coat a larger roller placed below the forcing rollers with a layer of filling. The thickness of the layer is adjustable, and a scraper ensures that the filling is removed from the forcing roller. As the wafer passes beneath the depositing roller, a scraper removes the filling from the roller and transfers it to the wafer. The table beneath the depositor is adjustable for height, and in cases where only a very light deposit is required, the wafer is brought close to or in contact with the depositor roller, and the depositor scraper is usually replaced by a taut wire. Creaming units can be fitted with more than one hopper, so that a variety of fillings may be included in one built-up sandwich.

Sandwich building may be a simple hand operation, or a more complex automatic system, where creamed wafers wait and build up on a specially designed unit. This unit controls the creaming unit so that every fourth wafer (of a 4-wafer sandwich) is not creamed. Automatic sandwich building can also be designed to run from a battery of wafer ovens. Thus, to make a 4-wafer sandwich, three ovens would be producing sheets for creaming, and one producing the top wafer. Each wafer proceeds and meets to form the sandwich. Once the sandwich is built, it passes beneath a roller of the predetermined height above the conveyor to ensure that the sandwich is evenly filled, and that each layer is adhering to its neighbour. This pressure helps to prevent subsequent separation during cutting and other operations. To ensure that the wafer sandwich is firm to meet the cutters, it passes through a cooling tunnel where a temperature of 7·8-10·0°C (45°-50°F) is desirable.

Wafer cutters cut a low stack of wafers by means of either wire cutters, rotary saws, or oscillating band saws. The cutters are spaced at the desired intervals, and the stack of wafers is pushed through them in a frame, either by hand or machine. The frame is withdrawn, transferring the wafers to a second frame which is then moved at right-angles to the first through a second set of spaced cutters, reducing the wafers to the required size, ready for packing or further processing.

CREAMING MACHINES

There are two basic designs for machines in use for sandwiching cream-filled biscuits. There is the high-speed track type, and the slower stencil-type machine. The high-speed track model may operate on one, two, or four tracks, with corresponding increases in production. Basically, biscuits are fed down inclined magazines, either direct from the oven cooling conveyor or from base tins, to pass along a chain conveyor which is narrow, to suit the biscuit dimensions. The base passes beneath, and almost in contact with a synchronised rotating stencil, which is situated in a tube through which cream is being pumped. The stencil is designed to fit the biscuit, and as the cream is extruded by the pump pressure a taut wire cuts off the filling, transferring it to the base biscuit. The amount of filling extruded is dependent upon, and regulated by, the pressure maintained by the pump. The filled biscuit passes to the second set of magazines for the top biscuit to be placed in position. The sandwiched biscuit is then pressed to ensure adhesion. It is straightened prior to the biscuit from each track being separated into two columns. These leave the machine stacked on edge, ready to pass through a cooler, or for feeding direct to the packing and/or wrapping machine. While this type of machine is fast and capable of producing a considerable weight of production, it is not as versatile as the slower stencil type. The high-speed machine requires a cream with rather more body than the standard filling, and can only handle one filling at a time. The versatility of the slower machine is demonstrated by the fact that it can handle any size or shape of biscuit, provided that the necessary changeover equipment is available. A jam-filling machine can be incorporated after creaming and before topping. The machine consists basically of a line of magazines from which a row of biscuits is fed on to the conveyor web (or apron) by means of a pusher plate. The rows pass beneath a specially designed stencil plate, and the table lifts the biscuits up to the underside of the stencil cut-outs. The cream hopper slides across the stencil, filling the cut-outs, and returns to its original position. As the table lowers, prickers ensure that the filled bases fall cleanly away with the web. The bases pass to the second set of magazines for the tops to be placed in position. They are then straightened prior to passing through the cooling tunnel, where the soft filling suitable for this process hardens to facilitate handling and packing. Control of the cream deposit is dependent upon the thickness of the stencil plate (and also on the specific gravity of the filling).

N.B.: The term 'cream' referring to the filling used in sandwich biscuits can lead to legal problems, and reference should always be made to the appropriate regulations in force.

If jam is to be included in the sandwich, an attachment can be fitted whereby jam under pressure is distributed through a compressed air controlled depositor directly on to the base after the filling cream has been stencilled, and prior to the top being applied. To ensure ease of running and that the jam sets, only good quality seedless jam should be used. While in use, the jam must be hot enough to flow easily, but not so hot that it melts the fat in the filling.

MARSHMALLOW EQUIPMENT

Marshmallow syrup is usually aerated either by a pressure whisk or by a continuous mixer. The type of depositor used depends upon the method of mixing. With a batch-type mixer, a gravity-fed-type depositor is used. The continuous mixer works as a pump, direct to the depositor head, without an intervening holding hopper. In both cases, pusher-bar-fed rows of bases from magazines approach the depositor head. In the case of batch process mallow, the table lifts the bases up beneath the depositor nozzles, which are available in many designs, to give various styles of deposit. The mallow must be of a suitable consistency, so that when the depositor plungers are raised, the suction draws the mallow down to be ejected when the plungers are lowered. To cause a clean break as the deposit is completed, the table lowers, and the filled bases move forward. A divided hopper, using two colours and flavours, can be used to provide further interest and variety. Jam or jelly can also be deposited with the mallow, timed so that it is completely covered by the mallow or so that it shows. The depositor, which is used with the continuous mixer, is also capable of handling two colours and flavours of marshmallow or a combination of jam or jelly and marshmallow. The depositing manifold has two passages, to accommodate two different materials (or both may carry the same), which lead to nozzles through which the deposit is made. A slider valve is placed between the mallow and the nozzles, which opens and closes as required to control the time and size of deposit. On this type of machine, the base and the web are not arrested at any stage. The manifold moves along with the base while it deposits and returns for the next row of bases, or where two manifolds are in operation, for alternate rows.

To avoid the need for two continuous mixers when requiring two colours of mallow, it is possible to manufacture one colour (white usually) at the mixer. The mallow is then divided into two streams. One passes through a specially designed colour mixing device, whereby colour is injected to the stream. The stream then passes through a small mixing unit and emerges as required, to proceed to the depositor.

Both types of depositor leave the marshmallow with a point or tip, which is normally flattened by means of a moist web passing over rollers under which the goods pass. The moisture prevents the mallow from sticking to the web, but can cause many problems if there is excess moisture transferred to the mallow, particularly if the goods are to be chocolate coated. Another ingenious device is the Oakes Detipping Roll, which is basically a driven refrigerated roller upon which atmospheric moisture condenses and freezes. The marshmallow goods pass under the roller, the tips are flattened without sticking to the ice and without moisture being transferred.

Marshmallow goods are sometimes dressed with pasteurised coconut, sugar, or other suitable decorative materials. These not only add to variety, but for uncoated mallow, aid handling and packing. Sprinklers, similar to those already outlined

(Chapter 19, 'Machine room equipment'), may be used of a capacity suitable for the large quantity of material involved, which may be of a coarse nature, and will also require care in handling, so that a minimum of damage occurs to the material. To save material lying on, and not adhering to the goods, forced air is directed over them to remove any surplus for recovery.

The surface of marshmallow being very moist and extremely adhesive, the goods must pass through a skinning tunnel to dry out the surface, to aid subsequent processing or handling. Low humidity warm air is blown over the goods as they pass through the tunnel, to carry away moisture. Care must be taken to ensure that the temperatures are not high enough to affect the stability of the foam, or to leave the mallow at too high a temperature for efficient enrobing.

CHOCOLATE HANDLING EQUIPMENT

The various problems of handling chocolate have been discussed in Chapter 8 (Chocolate and cocoa products), and it is with these in mind that handling equipment is designed. The main problems are tempering and weight control (viscosity), followed closely by cooling before packing. Equipment is available from the simplest manually controlled units to the sophisticated fully automatic plants. Chocolate can be processed in the factory from the raw materials, but it is usually obtained ready for use in blocks, in chip form, or in liquid form in bulk. If in bulk liquid form, thermostatically controlled storage tanks are necessary, with pumps to deliver to holding kettles in the chocolate room. (Owing to the fine degree of temperature control required where chocolate is being handled, it is usual to have a separate area set apart for chocolate, to minimise ambient temperature variations.) If using solid chocolate, storage should be in a cool, dry area, from where it is taken to the chocolate kettles. The kettles are thermostatically controlled, double-jacketed containers, with a stirring paddle which completely covers the kettle bottom and sides, to prevent the possible occurrence of localised overheating. The stirring action must be thorough, but sedate, to discourage air intake, since bubbles present extra problems. A pump is usually incorporated, but the chocolate could be removed by gravity feed. In both cases, heated pipes (double jacketed with temperature-controlled water circulating in the outer pipe) are used at all stages to transfer chocolate from point to point. We have already referred to the three systems of tempering, whereby chocolate is: (1) tempered by batches in kettles; (2) When the melted chocolate is constantly fed into a tempered batch as it is being used, or (3) the method of passing melted chocolate to an automatic tempering machine, situated separately, or as an integral part of the enrober. Automatic tempering is achieved by accurate temperature control of melted chocolate from a holding kettle as it passes through or over a cooling unit, and is then heated to working temperature. Very sophisticated units are available to ensure high standards of temper and viscosity. With trickle feeds to a tempered batch, there can be considerable variations in viscosity, corresponding to the proportion of solid fat crystals present at any one time.)

To coat biscuits, the pieces are fed by pusher bar from magazines, or, if of difficult sizes and shapes, by means of a cradle feed which gives a constant supply of biscuits, not in rows, to the enrober unit. This consists basically of a wire conveyor which carries the biscuits through a curtain of liquid chocolate flowing from a small holding tank. The surplus chocolate fills a shallow tray, beneath and in contact with the conveyor. Consequently, the biscuit is coated all over. The coated biscuit passes under a blower to remove unnecessary chocolate from the biscuit and, if required, imparts a 'ripple' pattern to the surface. All surplus chocolate falls through the conveyor to the enrober holding tank for recirculation. To assist in removing surplus chocolate and to give a good finish, the final part of the conveyor passes through a vibration section. Finally, the coated piece is transferred to a specially treated surface conveyor. At the point of transfer there is either an anti-tailing roller which removes surplus chocolate from the base to prevent a 'tail' or 'foot' appearing, or a turnover roller to invert half-coated biscuits (half-coated biscuits pass through the chocolate on the conveyor, but not through the curtain which has been diverted at the sides to fill the bottoming tray). The turnover roller can be designed to impart a pattern to the biscuit as it passes over it. The conveyor surface may simply be a plastic-coated web with an embossed pattern or the biscuit manufacturer's name included in the design, or it may be foil sheets or special metal plaques, suitable for carrying in overhead

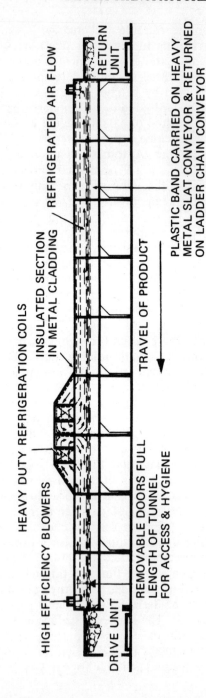

Figure 16. Line drawing of a Walden refrigerated air blast cooling tunnel

tiered coolers. (Cooling requirements are referred to in Chapter 8. See also Fig. 16.).

ICING EQUIPMENT

A bottoming, enrober type of machine, is suitable for icing biscuits, as it is usual to ice one side only. The unit consists of a hopper holding the icing, which is fed to a tray in which run two or three pick-up rollers. These lift the icing so that the wire conveyor carries the biscuits from pusher bar magazines through the icing. Sufficient distance must be given to allow excess icing to return to the tray before inverting on to the next web. At this stage, while the icing is still very soft, fine lines of a contrasting coloured icing may be piped on to form a pattern and added interest. This action also adds to the problems either of staff to decorate the biscuits or of supervision of a suitable machine. The base icing must be still soft so that the added icing blends with the base. A marbled effect can be achieved by passing the biscuit under trailing wires. Once iced and/or decorated, the biscuits must be dried out in a drying oven. The oven is in reality a conveyor system through a heated chamber. The conveyor is usually tiered to save space, and the temperature required is approximately 76-82°C (170-180°F) for about 1 hr.

CHAPTER 22

Ancillary equipment and automation

WHILE it is not possible to catalogue and describe all the equipment used in a biscuit factory, an attempt has been made to deal with the fundamental pieces of equipment in a general and basic manner. Some further items which, although outside the scope of previous chapters, are too important to pass over.

SUGAR GRINDING

The highest proportion of sugar used in a biscuit factory is icing sugar, but since it is usual to purchase only granulated sugar for general use, it is necessary to grind it to the icing form. For this purpose, a peg mill is probably the most suitable (these mills are also known as pin disc mills). Granulated sugar is placed in the feed hopper and passes by means of a vibratory feeder through a magnetic chute which extracts ferrous metals, if present. The sugar may pass directly into the central feed aperture by gravity, or may be lifted by suction to the mill when it is situated at a higher point than the feed hopper (suction feed reduces the production capacity considerably). The peg mill consists of a stator and a rotor, each with concentric rows of pegs which, when rotating, pass extremely closely to each other. The centrifugal force set up by the rotor, throws the sugar outwards between the rotating and static pegs, which reduce the particle size. The fineness of the grind obtained is determined primarily by the selection of grinding discs from a range with differing numbers of pegs and, secondarily, by the speed of the rotor. The ground material is collected by the force set up by the rotor, delivering it to a cyclone collector where the sugar falls to the bottom, where there is a bagging off or storage hopper outlet, and the air is drawn off at the top through a filter to retain the sugar. The air extraction tube should lead to an explosion chamber for safety.

Biscuit grinding

Scrap biscuits, if they are to be used again, usually require to be reduced to a meal or to crumbs. A peg mill can achieve this, but a simpler and less expensive machine is normally adequate. The simplest form is the kibbler, which consists of a hopper, at the base of which a removable disc is arranged horizontally. According to the disc selected, this has perforations of various sizes. Above the disc is a spring-loaded 'spider' (four-bladed, propeller-like rotor) which, when rotating, forces the biscuits through the apertures for collection down a chute. A variation on this type has perforated plates, which fit approximately one-third of the external circumference at the base of a cylinder. Inside the cylinder, a rotor forces the biscuits in a grating action against the perforated plates. This type of machine is also useful for sieving various materials.

OIL SPRAY MACHINES

Oil spray machines are used to impart a glossy finish to crackers, particularly of the savoury type. The application of oil tends to give an impression of richness to the cracker; aromatic flavours can be included in the oil to give added flavour or to create variety. Spraying with oil may, if care is not exercised, accelerate the onset of rancidity, and unless the percentage uptake of oil by the crackers is not strictly controlled, it can be a very expensive process.

Oil spray units consist basically of a thermostatically heated tank into which the oil or fat is placed. The heated oil can then be pumped under pressure through a filter and to one of two spraying outlets. The sprays are duplicated, so that when a blockage occurs, the spray can be switched over without loss of production. It is usual for sprays to be situated above and below the wire conveyor, which carries the crackers as they emerge from the oven band, so that the crackers are coated on both sides. The spray is produced by atomising the oil through nozzles. The spraying area is well covered to prevent oil-mist loss into the atmosphere, and surplus oil returns by gravity through a filter to the storage tank at the base. All pipelines are lagged and heated to ensure complete fluidity of the oil. A variation on the nozzle type of spray is to dripfeed the oil on to quickly rotating dispersers, which produce by centrifugal force a mist through which the biscuits pass. Although this system does not appear to be so positive as the previous type, it is unlikely that it will suffer from nozzle blockages so readily.

CONVEYORS

The problems of moving goods in the factory and the limitations imposed by restricted space, have been studied carefully by the manufacturers of conveying systems. Many systems are available—overhead mono-rail, gravity rollers, driven rollers; continuous webs, both wire and canvas; vibratory and air lift systems. The biggest problems surround the movement of biscuits in an unstacked condition (plain or processed), and the high rates of production in relationship to the lack of space. To ease this problem, webs (wire, canvas, and plastic coated) can now be made to turn through 90 and 180°. Units are made up in right-angle, quadrant, and half-circle turns or, if required, special angles can be prepared. Where space is particularly limited, wire conveyors can be arranged in spiral form to fit a long length into a relatively small volume, e.g. 1000 ft in a space 20 x 20 x 10 ft (300 m in 6 x 6 x 3 m). An additional advantage of the spiral form, is that the unit can be readily enclosed, and heating or cooling applied as required to process while conveying.

METAL DETECTORS

There are obvious reasons for wishing to exclude pieces of metal from biscuits and other foodstuffs. Firstly, to make sure consumers do not find metal in bought goods, as this can lead to discomfort or injury on the part of the customer and loss of sales and bad publicity on the part of the manufacturer—and even costly lawsuits. Other reasons are mainly concerned with risk of damage which a piece of metal can do if it passes unnoticed through expensive machinery. Rotary moulder rollers, blades, and webs, can be ruined. Pumps, brake and gauge rolls, cutters, and many other items of machinery, are similarly vulnerable. Pieces of metal can originate in raw materials or ill-maintained equipment—nuts and bolts, washers and screws from machines at all stages of production, pieces of wire from the sieves which are supposed to be removing such things. Even when all raw materials are carefully sieved and passed under electro-magnets, it is still possible for stray metal to remain or be introduced to the goods at one stage or other. Ferrous and non-ferrous metals can be detected by means of electronic

ANCILLARY EQUIPMENT

devices at strategic points in production lines to save damage occurring.

A metal detector consists of a search coil system, usually made of three coils and an electronic control unit, comprising an oscillator, an amplifier, and a control relay. The search coil arrangement consists of three coils wound around an aperture. The centre oscillator coil has a receiver pickup coil on each side of it. The oscillator coil sets up a high frequency field (like a radio transmitter) in the search coil system. The product being inspected goes through these three coils in turn as it passes through the inspection aperture.

It is arranged that under normal conditions, with no metal passing through the aperture, the search coil system is 'in balance'. That is to say, the field around each receiver coil is exactly the same, so the voltage induced in each receiver coil is also the same. These two pick-up coils are connected together in such a way that these two equal induced voltages cancel each other out and there is no resultant output signal.

A metal-free product going through the aperture, passes through the high frequency electrical field, but does not influence it. When a piece of metal, no matter how deeply it is embedded in the product, passes the aperture, it distorts the previously symmetrical high frequency electrical field, this distortion being first near the receiver coil and then near the receiver-2 coil. We thus have conditions where the field is unbalanced and the two receiver coils now being influenced by unequal electric fields, the voltage induced in each receiver coil is different and they do not cancel each other out. The resultant out of balance signal is fed into a high gain amplifier, where it is used to trigger the output control relay and actuate alarm or automatic ejection systems. (See Fig. 17).

The minimum size of metal that can be detected, depends upon the size of the aperture through which the product passes; the smaller the aperture, the smaller the minimum size that can be detected. Ferrous and other magnetic materials and alloys cause the greatest distortion to the high frequency electrical field because of their very high permeability, and are usually easy to detect. Non-ferrous metals, having high electrical conductivity, distort the high frequency field almost as much as ferrous metals, so metals such as copper and aluminium are equally easily detected. Non-magnetic stainless steels (18-8 austentic 300 series) are difficult to detect because they have

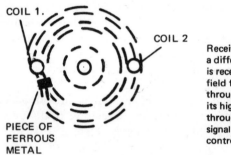

(a) Balanced field in search coil with no metal in the aperture

Each pick up coil is influenced equally by the balanced High Frequency field. No resultant signal to the amplifier.

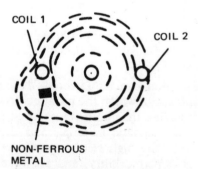

Receiver coil 1 is influenced by a different shape H.F. field than is receiver coil 2, because the field finds it easier to "crowd" through the ferrous metal with its high permeability than through the air. Out of balance signal is amplified and triggers control relay.

(b) Field distorted by piece of Ferrous metal near coil 1

Signals pulled up by receiver coils 1 and 2 do not balance out, resultant out of balance signal is amplified and triggers control relay.

(c) Field distorted away from the Non-Ferrous metal because of eddy currents induced in it

Figure 17. Diagrams illustrating how metals affect the electromagnetic field of an electronic metal detector.

very low permeability and high resistivity. Non-ferrous metals cause a disturbance in the high frequency field because of the eddy currents induced in them. The higher the resistivity of the metals, the lower the value of eddy currents induced on them, and the less the distortion produced in the HF field. Hence, the larger the minimum size of this type of metal that can be detected.

Detection sensitivity also depends upon the surface area of the metal rather than upon its volume. This is the reason for what is known as orientation effect, e.g. a long, thin piece of wire is easier to detect when it passes through the aperture orientated in one direction than in another. A metal sphere is often used as a test standard for metal detectors because (a) a sphere has no orientation effect, (b) a sphere, with its minimum surface area for a given volume, is the most difficult shape of metal to detect.

When selecting a metal detector system, detection sensitivity is important, but it is only one of a number of essential requirements. At the required operating detection sensitivity, the detector should be quick and easy to adjust by an unskilled person. Once adjusted, it should be quite stable, with no false alarms, and should not need readjusting for very long periods (several months). False alarms should not be set off by the normal vibration and electrical interference found in an industrial environment.*

The alarm system can function in one of several ways, depending upon the detector used and the type of material it is monitoring. If situated in a pipeline with liquids passing through, a valve system will close the accept valve and open the reject valve for a specified time. In the biscuit factory, detectors are usually placed on conveyors and the alarm system can:

(1) Stop the conveyor;
(2) sweep the goods off the web diagonally, by means of a plough ejector;
(3) open a trap—often a short slide prior to the stacker;
(4) mark the area with a brightly coloured plastic disc (or coloured sugar or flour);
(5) spray paint from an aerosol container, or a coloured dye by compressed air, on to the container;
(6) by means of compressed air, blow the suspect items off the web into a receptacle; or

---

* The author is indebted to Mr A. M. Lock, Managing Director of A. M. Lock & Co. Ltd, of Oldham, for this description of electronic metal detectors.

(7) push the packet off the conveyor by means of an air operated ram.

The search head and ejector system are synchronised with the web speed, so that the appropriate place is attended to. The restart can be automatic, manual, or remotely controlled, so that a person in authority can take the responsibility of starting up production again.

AUTOMATION

The production of biscuits enjoys a high degree of mechanisation, but complete automation has not yet been achieved. Before a plant can be controlled by computer, techniques of automatic sampling and testing have to be perfected, so that information can be fed to the computer which will make a decision upon the information received and take the appropriate action. To warrant the expense of automation, considerable savings in the cost of production must result from its installation. It is not difficult to accept that if automation can regulate production more reliably than manual control, enormous savings can be made if only by the reduction of 'give-away' or overweight on biscuit packs. Other savings are anticipated from a reduction in the labour force, and in more efficient stock control and planning, which should result from the use of a computer.

For computer control to be successful, it is necessary to monitor each part of the process, and for the computer to be able to correct as errors occur. Initially, all raw materials will need expert testing, so that variations can be allowed for at the appropriate stage. The mixing technique and dough consistency must be within fine tolerances. Present investigations by the Flour Milling and Baking Research Association and the British Scientific Instrument Research Association, have found a close relationship between dough consistency, power input, and dough temperature, when mixing semi-sweet, hard dough biscuits. As a result, techniques and instrumentation are being developed to control dough consistency automatically. Control of the dough sheet is well advanced by continuously measuring its thickness, and automatically controlling the gap and speed of the appropriate gauge roll and web, so that a consistent thickness is presented at the cutter.

ANCILLARY EQUIPMENT

Oven control depends upon sensitive thermostats, coupled with the fuel supply and positive 'on/off' firing, with a wide, but controllable, variation in the degree of heat produced. Zone conditions require pre-setting, and when variations are needed, manual control is still necessary. Research is being pursued into determining and controlling humidity in the baking chamber.

Assessment or control of biscuits, when baked, is well advanced, and devices are available to measure automatically the colour, diameter, stack height and weight of biscuits. Biscuit colour is assessed by the amount of light reflected from the surface of the biscuit, compared with the amount reflected from a selected reference surface. The Biscuit Sampling and Automatic Measuring Equipment (known as SAM for short), measures the length of a row of 10 biscuits placed edge to edge and then the stack height of 26 biscuits. Both measurements are made by a light beam falling on to photodiodes. They are accurate to 0·1 in (2·5 mm) or less. The column of biscuits is weighed prior to being returned to the main stream of biscuits on the conveyor. Determination of moisture content is semi-automatic, by means of dielectric heater electrodes being mounted above and below a weighing balance tray upon which the sample of biscuits is placed. The instrument weighs the biscuits before and after drying and prints the weights automatically; normal drying time is 5 min. The final control is of the packet weight as it passes over a check weigher.

All the controls are connected to data loggers, and a complete record of all variations is maintained. Wherever control of this nature is applied, the information can be fed to the computer which, according to its programme, can take the necessary steps to correct errors beyond the acceptable tolerances. For example, if the biscuit colour becomes excessive, the computer will reduce the heat in the oven. If the moisture content is too high and there is insufficient colour, the heat will be increased.(It may be necessary initially to increase the baking time and, if this is the case, it will automatically slow down the cutting machine and gauge rolls as all the units are fully synchronised. When the oven temperature has built up again, the baking time will be corrected along with the machine speeds.)

CHAPTER 23
Packaging of biscuits

AS biscuits are both fragile, and highly hygroscopic, they require protection from damage and from atmospheric spoilage. Historically, biscuits have been packed mainly in not quite square tins which hold about 8 lb biscuits. The rectangular shape permitted a wide range of biscuit sizes and shapes to be economically packed to a fairly standard weight. The biscuit tin was very convenient for packing, storage and transport, but the biscuits suffered in the shopkeepers' hands when the lid was removed for serving or display purposes. The cost of the tin was quite low per pound of biscuits if re-used (after thorough washing and drying) for several journeys. However, with the development of more sophisticated packaging materials and machines, and the change in the consumer's shopping methods, the biscuit tin in its 8-lb form is no longer frequently used. The prepacked unit of 6-8 oz is now the general rule. Tins are still widely used for speciality packs on a non-returnable basis, and these prove very popular as gifts in a wide range of sizes and designs, at Christmas time in particular.

The 8-lb biscuit tin fulfilled its requirements satisfactorily, but with the introduction of self-service grocery stores, the tin became obsolete, and marketing of biscuits entered a new phase. The biscuits had to appeal to a new customer in a new environment. The housewife, among whom many work full-time, is pressed for time, and in the self-service stores passes the biscuit shelves in two or three seconds. In this short time, the packet of biscuits must attract her eye and persuade her to place it in the basket. This one factor has alone led to a revolution in display and design. Whatever the quality of the product, if the wrapper does not attract and convince the customer that the contents are of high quality at a price within her range, and that they are manufactured by a company in whom she has confidence, then the packet will remain on the shelf. This is a complex message to convey in three seconds and much research has been done to find how to achieve this end. The use of transparent wrapping films to show contents, of colour, printing, and the development of individual and easily recognised 'house' symbols, are all techniques widely recognised as ways to attract the shopper.

The wrapper must, of course, fulfil other requirements, the chief of which is protection. Firstly, the fragile contents must withstand considerable handling by the packing, despatch, and transport departments of the manufacturers; by the retailer, and by the customer; and be in good condition when about to be eaten. The wrapper, therefore, must protect against breakage. Secondly, it must also protect the contents from contamination by dust, dirt, strong aromas, bacteria, and mould. Any one of these can arise during transit and storage and while on display at the point of sale. The wrapper should protect the contents from infestation by flies and other insects. The wrapper's third role, is to protect the contents from deterioration and spoilage, particularly from an increase in moisture content of the biscuits. Biscuits being liable to problems of rancidity also, the wrapper should act as a grease barrier, for fat which has seeped from the biscuit into the wrapper will readily turn rancid. Daylight and oxygen will accelerate the onset of rancidity, and the wrapper should also protect the contents from these. It is desirable that the wrapper should continue to give protection to the biscuits after the housewife has opened the packet. Few housewives store the biscuits in an air-tight tin after opening the packet, and as the condition of the biscuits rapidly deteriorates after opening, unless the wrapper can be effectively re-sealed, then the manufacturer will be unfairly blamed for low quality products.

Packaging biscuits in predetermined weights has reduced the problem of broken biscuits from the grocer (the customer now receives the broken-end biscuits in the packet). The grocer also benefits from selling a certain number of packets, rather than giving excess weight each time he weighs the biscuits from an 8-lb tin.

There are, of course, contradictory requirements when considering the desirable qualities of a wrapping material. Transparency, to show the goods inside the packet, will not exclude daylight from the goods, consequently, certain compromises have to be made. Further problems and compromises arise when consideration is given to the available wrapping materials and to their relative costs.

WRAPPING MATERIALS AVAILABLE

Various forms of paper materials are used for wrapping and

packaging biscuits, the most widely used is greaseproof. This is a paper developed for wrapping butter and fats, and because of its ability to form a barrier against fat (and daylight), was originally used in 8-lb tins, but is still used in packets. Greaseproof is available in a corrugated form, which affords some protection against physical damage. Corrugated greaseproof is also used to form skillets, or trays, into which biscuits are placed before being over wrapped. Glassine, which is a translucent, glazed greaseproof, is available in a wide range of strong attractive colours. It is widely used in speciality packs and in conjunction with chocolate finished goods. It is impervious to grease and, if chocolate coloured, does not show chocolate markings. Other colours are used for contrast and aesthetic appeal. Greaseproof and glassine are available as paper shavings, which are used as packing in spaces in speciality packs. Waxed paper is paper with a wax coating on one or both sides to give moisture-proofing and heat sealing qualities. Cardboard is used to form skillets, trays, and cartons, but must have a grease barrier between it and the biscuits to prevent fat migration. Generally, a greaseproof paper liner is used either inserted loosely or laminated to the cardboard with wax. The barrier may also be formed by coating the cardboard with polythene or similar greaseproof material, or by the insertion of a vacuum-formed plastics tray. This final method is being widely accepted for assortment packs, as the shapes of the various biscuits can be defined in the tray and consequently, act as a guide to the packer.

Direct wrapping of biscuits in a plain impervious protective wrapper, permits insertion of this pack into a printed chipboard carton. This method produces a relatively low-priced packet, with a reasonable, but not high quality, printing surface, which is readily handled by machine and gives adequate protection to the contents at all stages prior to purchase.

Transparent cellulose film, popularly known as 'Cellophane' (the registered trade mark of British Cellophane Ltd) is very widely used for wrapping purposes. It is available in different types and has different properties. The uncoated PT-type film is highly permeable to moisture and air, but is dust-proof and greaseproof, it is therefore only of use around already moisture-proof wrapped packets or containers. It is not heat-sealable. When coated on both sides with a nitro-cellulose lacquer, the regenerated cellulose film is known as MSAT. It has considerable resistance to moisture and oxygen, and can be heat sealed. The main disadvantage of MSAT is that it contracts on losing moisture (and expands on gaining moisture). As biscuits are more hygroscopic than the film the latter will lose moisture to the former, thus contracting and tightening the wrap. Sometimes the stress may be sufficient to cause the film to split. Under dry or cold conditions (below freezing), the film becomes very brittle, but can be softened by the use of polyalcohol plasticisers which increase its flexibility. The coating lacquer is to some extent damaged when in contact with printing, and consequently, the moisture permeability will be increased. MXXT cellulose films are high-barrier coated films, but the nitro-cellulose lacquer is replaced by polyvinylidene chloride co-polymer (PVDC). Two types are available, depending upon the coating process, MXXT/A and MXXT/S, MXXT/A having superior barrier properties to MXXT/S. The co-polymer coating is tougher than the nitro-cellulose lacquer and can withstand printing and creasing without increasing its permeability. MXXT is less permeable in both forms to moisture than MSAT, and less liable to dimensional changes as the humidity alters.

Paper, such as glassine and bleached Kraft, can be coated with co-polymer to improve resistance to moisture and to make the pack heat sealable. Several coatings are necessary on the inside of the paper to ensure that none of the fibres protrude through and contact the biscuits. Otherwise fat would migrate into the paper. Only a light coating is given to the outside, usually over the printing, to improve the gloss and the heat sealing qualities.

Polythene film is not a very satisfactory wrapping material for biscuits by itself because, although it is satisfactorily resistant to moisture, it is a very poor oxygen and odour barrier. Polythene tends to be slightly 'milky' in appearance (it lacks sparkling clarity of cellulose film) and requires special heat sealing equipment. It is, however, very useful in conjunction with other materials, e.g. coating on cardboard skillets and as a laminate with cellulose film. It is also used as liners for outer containers and for shrink-wrapping several units together.

Polypropylene film, when coated on both sides with co-polymer (PVDC), is a satisfactory wrapping film when the temperature control of the heat sealing equipment is sufficiently precise, although its running performance on packaging machinery is inferior to that of cellulose film. Owing to its rigidity, very thin gauges can be used, and its relatively high price per pound is offset by its high yield in area, and in fact compares well with MXXT.

Aluminium foil is an exceptionally good moisture, fat, and odour barrier, but is expensive and usually has pin holes in it. On its own, it is rarely used for biscuits, but is often lacquered for heat sealing with nitro-cellulose, or extrusion coated or laminated to polythene or polypropylene. This coating gives an exceptionally fine finish and protection, suitable for high value chocolate count lines, and speciality high quality and export packs. Foil is also laminated with wax to tissue, greaseproof paper, or cellulose film.

Many laminated wrapping materials are available, based on the previous materials, with many combinations and generally high qualities. For example, two layers of cellulose film (MSAT or MXXT) can be laminated together with wax. This will give improved protection and strength. If the upper layer of film is reverse printed, the appearance of the wrapping will be excellent, and the lower film will protect the biscuits from the printing ink which, even if not dangerous to health, could accelerate rancidity if in contact with the biscuits. When tissue paper is laminated with wax to greaseproof paper, a good protective layer is formed, and when heated, the wax permeates the tissue to give heat sealing properties.

A recent development in wrapping materials is metallised cellulose film. Aluminium is vaporised under vacuum by the intense heat of an electric arc. The vapour condenses on the MXXT/A film and, although no claims are made for a reduction in permeability, a very high quality material is formed with high resistance to creasing, and offering a good barrier to light.

Irrespective of the wrapping materials, resistances to moisture, oxygen, and odours, these can be relegated to insignificance if the material will not withstand folding or creasing, if the printing has impaired the coating and, most important, if the seal is not satisfactory. The quality of the film depends upon the quality of the seal. Sealing is dependent upon the type of coating, the temperature of the sealing jaws, the pressure of the jaws, and the dwell time of the jaws. All these factors are variable, according to the type of wrapping material, the type of seal, and the type of wrapping machine involved. MSAT-type cellulose film gives optimum sealing strength from

120-130°C (248-266°F), with an adequate seal being obtained ± 10°C (± 18°F), MXXT/S has a wide temperature tolerance from 90-190°C (194-374°F). While MXXT/A has a narrower range 150-190°C (302-374°F). It gives a much stronger seal. Coated polypropylene reverts (permanently shrinks) at 150°C (302°F), and although there are variations of temperature requirements from different manufacturers, a sealing temperature of 140°C (284°F) is considered necessary with a range of only ± 5°C (± 9°F). [British Cellophane Ltd claim a sealing range of 120-140°C (248-284°F)]

The relative humidity of the stores can affect the moisture content of cellulose films, leading to further variation of sealing qualities, as the heat has to drive away moisture before reaching the required sealing temperature. Storage conditions are particularly important for wrapping materials, and a recommended temperature of 20°C (68°F) with a relative humidity of 40%, should be maintained. Reels should be brought to the working area 24 hr before use. Dimensionally unstable materials, which expand as the moisture content increases, will tighten on the reel, and wrinkling will occur if the humidity is high. The reel will not run smoothly on the machine and, in excessive cases, low temperature sealing will prevent the film from running at all.

Other factors which affect running performance are the quality of the finish of the material, the rigidity of the material, and the method of feed by the machine. The 'slip' quality is mainly dependent upon the polish on the surface of the material. Rigidity varies with each material and also with the gauge or thickness. If materials are laminated, the rigidity and gauge are usually increased. The wrapping materials based on paper and on foil are normally rigid, as also is polypropylene. 'Push fed' wrapping machines require rigid materials for best performances and, if low-rigidity materials are to be used, a positive mechanical feed is necessary to keep the material running without folding.

WRAPPING MACHINERY AND SYSTEMS

There are a perplexing variety of wrapping machines used in the biscuit industry, ranging from slow-speed models, wrapping 20-40 packets per min, to high-speed models, capable of wrapping over 350 items per min. The machines are designed for handling biscuits in many different ways, from direct wrapping and over-wrapping, to tumble packs.

The most popular form of packet is the direct wrap of a column of biscuits standing vertically on their edges. These can be fed direct from the cooling conveyor into the pockets of a continuous conveyor. The length of pocket controls the number of biscuits which can be wrapped as a packet. The packet weight is consequently controlled by the number of biscuits present. If the biscuit size varies beyond certain tolerances, the packet weight will increase or decrease accordingly, and in either case many problems arise. Many attempts have been made to control weight variations, but it is well known that biscuit weights can vary considerably across the band as well as along the band. The monitoring systems and precision machines, referred to previously, reduce this problem considerably, but until they can be widely introduced, simpler systems are used. In an effort to reduce the variation between rows of biscuits, two methods have been developed to mix biscuits from different rows. The simplest, is to bring three or four rows together through a zig-zag channel, placed on an accelerated web prior to stacking. The acceleration enables the biscuits to spread out and merge without piling up. Once the biscuits are stacked, they proceed in, say, four rows instead of twelve or sixteen to four wrapping machines, which handle the entire oven production. A rather more complex system depends upon the single rows being fed automatically from the cooling conveyor into a series of magazines, from which a pusher plate feeds one from each magazine, giving a complete blend of biscuits from across the band instead of from along the band. If the number of magazines exceeds the number of biscuits required for one packet, the pusher plate can select the number, and so that one or more magazines are not under-used, the pusher plate is so arranged that it misses successive magazines each time it feeds. The biscuits are then stacked as necessary and wrapped.

Wrapping machines are usually designed to wrap one shape of biscuit and can usually be adjusted to accommodate different sizes of that shape. Some machines are capable of being adjusted to wrap all shapes of various dimensions. Direct wrapping machines are available to wrap biscuits standing flat in low piles, either singly or in two or more piles. Most machines

PACKAGING OF BISCUITS

can double wrap the biscuits (whether in piles or in column form), using, for example, two layers of heat seal material, or one layer of heat seal over a layer of non-heat seal material (such as greaseproof paper or, for fragile biscuits, a heat seal material over a layer of corrugated greaseproof paper). The wrapping material can be sealed with double point end folds and/or with end seals of a heat seal material. The longitudinal seal may be by the most economical and simple overlap, whereby the two ends of the wrapping material overlap each other and heat is applied along its length. This is not so effective as the fin seal, where the two ends of the material are brought with the inside surfaces together. The heat sealing jaws exert pressure on each other, making an excellent seal, and no pressure or heat is applied to the biscuits, and this helps to reduce breakage. The fin seal will normally lie flat against the biscuit packet as the end folds draw it close to the packet. A further variation in longitudinal seals is the double fold, whereby the two ends are brought together with inside surfaces touching. However, one end is longer than the other, and the long one is doubled over and folded behind the short end, completely enclosing it to make a very good seal.

So that the wrapper design registers correctly with the packet, a photo-electric cell is incorporated on the machine which is synchronised with a mark on the wrapper. Tear tapes can usually be added to the wrapper during the wrapping action, to facilitate packet opening by the customer. Date coding is also available if required.

Direct wrapping machines are often of a relatively lightweight nature and are mounted on castors so that they can be moved to the point at which they are required to work, and moved away for changeovers, being replaced by a similar machine which has been previously prepared.

High-speed machines (capable of up to 350 wraps per min) require either a fairly rigid item to direct wrap or, if the item lacks rigidity or strength, it needs to be packed first in a tray or skillet. Over-wrapping is then in a pillow pack. The pillow-type pack is made by forming a tube around the product. The longitudinal seal is fin sealed and the tube is crimp sealed at each end of the pack. A knife cuts the seal across the middle, so that each seal forms the final seal of one pack and the first seal of the second pack. Instead of cutting, the packs may be left in strips of a definite number, perforated for easy separation. The base seal lies flat as the end seals retain it in position.

There are various machines available to erect blanks of card or corrugated paper into cartons or skillets, and these are capable of very high speeds. Often it is sufficient to have a carton-making machine connected directly to an over-wrapping machine, so consequently, similar operating speeds are desirable. If gluing of the cartons is necessary, an edible grade glue should be used, and preferably one which makes a fast firm bond and dries rapidly. Carton-making machines are also capable of filling them with pre-packaged materials, which are then sealed in.

Collating and wrapping machines form a bundle or stack of packets and then overwrap them. Often the shrink type of material is used for this purpose, as it holds the items firmly together. Machines will also collate packets and feed them into cartons, which pass through a gluing machine to seal, prior to despatch.

Particularly useful for savoury and small snack-type biscuits, are the pillow-type tumble packs. These are formed around the filling tube of a funnel into which is tipped, from a bucket feed, the weight or volume of biscuits for one packet. As the wrapping material is formed into a tube, it is sealed by a simple overlap or, if necessary a fin seal. Top and bottom seals are crimp, and each pack is separated by a cut-off across the seal. Compressed air is usually injected to inflate the pack, to protect the contents against breakage. An inert gas, such as carbon dioxide, can be used if desired to delay oxidation, but this should not be necessary.

One of the most interesting developments in packaging has been the introduction of machines capable of transforming plastics materials into specially shaped trays for packing biscuits. These not only form an impervious barrier and give protective strength to packs, but when designed for assorted packs, save considerable time in packing, and in training operatives, since the appropriate place for each type of biscuit is distinct, as the moulded shape is so designed that only the correct biscuit will fit.

PART V
General considerations

CHAPTER 24

Factory layout and hygiene

FACTORY LAYOUT

There are so many imponderable factors concerning factory layout and hygiene that only the basic principles can be discussed. The types of products, the variety of plant, the degree of automation, bulk handling, the mode of despatch, the amount of capital available, and the provision for future expansion, are some of the incalculable factors affecting the requirements and limitations of factory layout. Existing buildings also influence planning, but whatever the beginnings and the ends, certain basic principles must be observed. Superficially, hygiene may have little to do with layout, but the biscuit manufacturer has an inescapable duty to the public to produce biscuits that are not only free from dirt, but are also sanitary and free from infection. The two are not synonymous, but are frequently closely related, and factory layout should be planned with hygiene as well as production and cost in mind.

The first consideration is, of course, the site. The main site requirements are the provision of water, gas, electricity, and the means for sewage disposal. It should be readily accessible, not only for deliveries and despatch, but also for the labour to work as operatives. It may not be necessary for there to be a large pool of labour adjacent to the site, but there should be adequate and convenient public transport available. From the hygiene point of view, it is no good being hygiene conscious inside the factory if the site is in an area where the atmosphere is filled with smoke and dirt, and if the surroundings are ideal breeding-grounds for rats, mice, flies, and other pests of insanitary habits.

In considering the factory layout, the basic principles are straight-line flow systems, minimum handling, and the shortest possible lines of communication. These principles are not always completely compatible, and although it may be necessary to compromise, the extent of the compromise frequently depends upon available capital. Planning the layout can be quite complicated, but can be facilitated by the use of models made to scale. The models do not need to be replicas of the plant or premises, but should represent them accurately regarding area and volume and should, of course, all be made to the same scale. In this way, at very little cost, some impression of the completed layout can be gained. By using three-dimensional models, the amount of headroom can be established, which is not always easy to visualise from drawings on paper. One other advantage of a model, is that plant can be arranged in various positions until the most satisfactory layout is obtained. This scheme is invaluable, not only for planning complete layouts, but also when planning expansion to existing plant and premises. Future expansion is a point to bear in mind when planning the original site and layout. Whether expansion is to be either horizontal or vertical, it is cheaper to prepare for it at the start, than to have major structural alterations at a later date.

The actual factory layout will depend upon production requirements. The size and type of plant will govern the size and extent of the production department, and will, to some extent, decide the requirements and positioning of the ancillary departments and offices. Bearing in mind the principles of in-line production, minimum handling, and short lines of communication, provision must be made for stockrooms of raw materials, packing materials, and finished stock; for a mixing department, if it is to be separate from the production department; for packing sections, including assortment packing; and processing departments and a despatch department. The stockrooms, particularly raw material and packing material stores, should have access to the departments with which they are concerned, should have temperature and humidity control, and covered off-loading areas. (When stockroom conditions are not controlled, it should be situated to the north of the factory to maintain the least variation and the lowest temperature.) The methods of off-loading will depend upon the situation and upon the type and quantity of stores to be handled. Bulk handling of raw materials presents few off-loading problems, but handling of unit packs can present many difficulties, chiefly concerned with the amount of labour required and the time necessary for different types of goods. The easiest and fastest method of off-loading, is by fork-lift truck from delivery vehicles, with the goods already palletised. Pallets are then directly stacked, with no further man-handling. The main drawback of this method, is the standardisation of pallets between the various delivery companies and the receiving factory. Short of this method, a conveyor belt system direct from the vehicle to the point of storage, or to a central point for stacking on pallets (which can then be taken by truck to the place of storage), is reasonably quick and requires only a small labour force. A completely manual system has the advantage of greater flexibility, but the speed of the operation is directly proportional to the size and efficiency of the labour force. The efficiency of the labour will depend frequently, upon the minimum requirements of a team on a particular off-loading job. Numbers above the minimum team strength will increase the speed, but probably not proportionately, while numbers below the minimum will completely upset the balance and sequence of operations, and efficiency will suffer disproportionately.

A mixing room, separate from the production departments, helps to localise flour and sugar dust, but tends to isolate, and consequently, communications may suffer unless a satisfactory system is operating. When fermented doughs are to be produced, it is desirable that temperature and humidity controlled conditions can be provided.

Similarly, isolating the machine room from the baking department, facilitates temperature control throughout the production department. In the processing department, it is necessary to provide different temperature and humidity conditions in the enrobing and in the packing areas, and so the sections must be separate from each other.

All packing sections should be either adjacent to the packing materials stores or should be fed with a supply of packing materials directly by some form of conveyor. The packing materials may be tins or cartons, for loose biscuits, or biscuits for processing or biscuits in packets. The tins or cartons will require preparation before being fed to the packing area, and tins have to be washed and dried previously. It is desirable to have direct contact between the tin preparation and the tin wash, which, in turn, is connected to an empty tin off-loading section. The filled containers should be conveyed from the packing section to a labelling section and then to finished stock or to the second process department. The movement of all biscuits, to whatever destination, should be either automatic or by pallet and truck. The use of overhead monorail systems can be useful in this concept, as a constant supply of various packing forms can be kept supplied to the packing areas, while the filled packs can be removed continuously and conveyed, via labelling, and second process departments, to the finished stock department and back to the packing material section (the pack being removed at the necessary stage).

Wherever palletised handling by trucks is employed, floors should be level and doors wide enough and preferably self-opening (or heavy duty rubber to withstand being pushed through) for the passage of the truck and loaded pallet. When handling is necessary on different floor levels, trucks should either be provided on each level, or lifts provided of such capacity to take a truck and loaded pallet.

The despatch department should be large enough to carry a comprehensive working stock from the finished stock department to enable van loads to be prepared in advance. The prepared loads should then be loaded directly either on pallets by truck, or wheeled skids or cages, or individually by a mobile conveyor reaching into the van.

The ancillary departments and offices should be situated in convenient positions, depending upon their functions. The departmental offices should be in the appropriate departments and the administrative offices together, but adjacent to the more important departments with which they are concerned. The laboratory is best situated close to the production department, but as it is concerned with most sections of the factory, this is not important. The maintenance department is usually a well-developed section requiring plentiful space and, although it should be near to the production department, it should be completely disconnected from any section storing, preparing, or handling raw materials or finished products.

Separate cloakroom facilities for both male and female operatives should be provided close to the entrance to the factory, then there is no necessity for the operatives to pass through the factory in their outdoor or ordinary clothes. Protective clothing should be put on in the cloakroom and worn at all times in the factory, and should not be worn at any time outside the factory. Apart from facilities for clothes and changing, cloakrooms should provide adequate toilet, wash basin, and drying facilities. A first-aid or welfare room should also be provided and, like the cloakrooms, should not have direct access to any sections storing, preparing, or handling foodstuffs.

FACTORY LAYOUT

Materials and general considerations

When considering the types of materials to be used, emphasis should be placed on what is required of them, their ability to be kept clean, and their general appearance. Floors should be hard wearing, easily cleaned, and should *look* as though they have been cleaned. The hard-wearing factor will vary, according to the position and type of traffic over each particular section. Concrete is probably the cheapest floor material, but rarely looks clean and is only hard-wearing to light or rubber-tyred traffic. Steel wheels and mixing tubs wear the surface and then break it up. Concrete floors are generally very dusty, but this can be prevented by applying a sealing fluid when the floor is laid. Where water or grease is liable to be spilt on a sealed concrete floor, it can be very dangerous. Concrete floors are only really satisfactory in stockrooms and other non-production department areas, where the traffic is light and the floor surface has been sealed. The most attractive floor surface is one of kiln-fired quarry tiles, particularly red tiles. These are readily cleaned and look well when clean. They are also relatively hard-wearing, will withstand normal traffic, but not mixing-room traffic, i.e. full mixing tubs with small steel wheels. They can be used to advantage in all departments, except those with the heaviest traffic. Steel 'tiles' are available for those section which will withstand the heaviest traffic, but even they will become damaged where the heaviest traffic is concentrated. Steel tiles clean easily, but are not as attractive as quarry tiles. Woodblock floors are quite serviceable, except where liable to contact with water or grease, and they should not be polished as they can be dangerously slippery.

Walls should be smooth and impervious, and where they meet the floor should be coved to eliminate crevices where dirt can lodge and insects and micro-organisms breed. The most attractive, durable, and hygienic finish is again, tiles, either ceramic or plastic, but unfortunately, these are the most expensive. At the minimum, half-tiling is worthwhile in the mixing room, otherwise good quality gloss or egg-shell paints, on a smooth plaster surface, are adequate. Distemper and lime-washed walls, although permitted, are unsatisfactory as they will not withstand washing and require fresh application at least twice a year.

Ceilings should be coved to the wall, and can be treated in a similar manner to the walls. False ceilings should not be fitted to blank off unsightly girders and pipelines, as this serves only to form nurseries for insects and rodents, out of sight and undisturbed. Ceilings suffer most above oven exits and where there are steam discharges. To minimise the difficulties, canopies with extractor fans should be sited above these places, to carry the waste hot gases away into the atmosphere.

In addition to considering layout and hygiene during planning, some thought should be given to the conditions of work and the reduction of accidents. Fatigue, leading to carelessness, is one of the greatest causes of accidents in factories, so conditions inducing fatigue in the operatives should be avoided as far as possible. Many of the factors considered to diminish fatigue are also closely allied to the principles of hygiene. In biscuit factories, where baking and similar processes involving high temperatures are carried out, it is important to instal an efficient system of ventilation. The opening of windows, although providing ventilation, is not very satisfactory, as the window permits flying insects and birds to enter, as well as dust and soot, and draughts may be caused which can be very damaging to fermenting doughs, chocolate, and biscuits susceptible to checking. Ideally, a thermostatically controlled system, bringing cleaned washed air into the factory and discharging the stale air to the atmosphere, can be used to maintain a constant temperature, cooling in summer and heating, when necessary, in winter. Lighting should be even and adequate throughout the factory. The maximum use should be made of daylight, preferably from the north, and excluding direct sun's rays. Artificial lighting should not be concentrated in the places of high activity, as it is as fatiguing to the eyes to become accustomed to changes of light strength, as it is to work in a poor light. Noise is a factor, frequently overlooked, which readily induces fatigue, and although noise cannot be avoided completely, much can be done to reduce it. Sections which are particularly noisy can be partitioned off with soundproofing materials; regular maintenance on the machines and routine overhauls to replace worn parts, even applying sound deadening materials to panels on noisy machines, are all factors to be considered to reduce noise in factories. The use of colour schemes in the factory can play their part in reducing fatigue and facilitating the application of the principles of hygiene through helping to improve the working conditions and the contentment of the operatives.

HYGIENE

The observance of hygiene is neither a fetish nor a luxury, its non-observance can be very expensive both financially and morally. Hygiene is concerned with preserving or maintaining health and, as producers of foodstuffs, the biscuit manufacturer is obliged to observe and honour the principles of hygiene. These principles are not concerned only with keeping premises and plant clean and presentable, but also with the prevention of micro-organisms, insect and animal pests, or other impurities from developing in or around the premises, and their coming in contact with the raw materials, finished products, and packing. Such contact could, of course, contaminate them, so as to make them dirty or insanitary. In this context, dirt refers to any particle that is repulsive or unpleasant in association with foodstuffs, whilst insanitary infers that the product could cause disease or be injurious to health.

Bacterial contamination

Contamination of an insanitary nature is normally due to the presence of micro-organisms, which may result from contact with operatives, insect and animal pests, or from insanitary premises. Chemical contaminants of a poisonous nature are also injurious to health. Micro-organisms fall into categories of bacteria, yeasts, and moulds. Although these groups thrive in conditions of warmth, moisture, and food, they have their own particular limitations, but it can be seen that bakery and chocolate room conditions are very suitable for their development and growth. All of them can cause fermentation or breakdown in some form of various raw materials and finished products, but the bacteria are the group associated with causing the breakdown of health in the human body. Certain types of bacteria are able to form spores which can withstand adverse conditions, or, they can be in the active form where they multiply by simple division when in conditions favourable to them. The bacteria feed on foods in solution in a similar manner to yeast, but often the waste products are poisonous to man and consequently, cause ill-health or even death. The waste-products may already be present in the foodstuffs, or the bacteria may be present and become active while in the digestive tract.

Bacteria (and yeasts and moulds) can be killed by the application of heat, but the time and temperature necessary depend largely on the type of bacteria and also whether present in the spore form, as the spore is usually more heat resistant. Very few bacteria are likely to survive the low moisture content and baking temperatures employed in biscuit making, but cream fillings, icings, coatings, and marshmallows undergo no such heat treatment and consequently here, the dangers are much greater. Marshmallow, in particular, also has a high moisture content, making it even more suited to bacterial development.

Bacterial food poisoning usually results from either contamination with bacteria of the salmonella group, or with pathogenic staphylococci. Food poisoning can follow from salmonella contamination through the bacteria being present when consumed, or through only toxic waste products remaining after heat treatment has killed the bacteria or through both causes. It can be seen, therefore, that killing the bacteria is insufficient. Development must be prevented. Salmonella poisoning occurs, after varying periods, from the time of consumption of the contaminated food, depending upon the degree of toxins present in the food when eaten. Pain occurs in the abdomen, which may be followed by prostration and violent vomiting and sometimes even by death. Diarrhoea, dysentery, enteric fever, headaches, rigors, fevers, and rashes, are all conditions which may be caused by bacteria of the salmonella group. Salmonella bacteria are conveyed to raw materials and baked products in a number of ways, but most commonly by 'carriers'. Carriers are humans or animals infected with the bacteria. Although they may not be suffering from any ill effects, the bacteria pass from the body in the faeces, from which they are transferred to the hands and then to the foodstuffs, unless careful personal hygiene is practised. Contamination from domesticated animals can be direct or through handling the pets; and from rodents, birds, insects, and flies, direct contamination from feet and faeces. Dried egg and dried albumen, duck eggs, and desiccated coconut, are frequently infected with salmonella bacteria. It is wisest, therefore, not to use duck eggs for any purpose in the bakery, and the other ingredients should be pasteurised before use, even when they are included in baked products.

Staphylococcal food poisoning, generally originates in infection from infected noses, throats, and respiratory tracts, and from the pus of boils, pimples, carbuncles, and suppurating wounds. Bacteria of this type, can also be present on mixing bowls and utensils. The poisoning is caused by the toxic waste products of the bacteria, which accumulate in infected foodstuffs of a suitable nature for bacterial growth, over a sufficiently long period. The most vulnerable media are milk products, creams, and meat fillings, so that this type of food poisoning is more likely to result from kitchens and flour confectionery bakeries, rather than biscuit factories, but that does not preclude the possibility. Headache, nausea and vomiting, diarrhoea, prostration, or even collapse, are the usual symptoms which may result from staphylococcal infection. Fortunately, death is a rare result. To prevent this type of outbreak occurring, operatives suffering from respiratory infections, boils, or septic wounds, should not be permitted to handle food. All operatives should observe the principles of careful handwashing and personal cleanliness. Foods that are susceptible to staphylococcal infection, should be handled as little as possible, and should be kept at refrigerated temperatures.

Chemical contamination

Other forms of contamination, liable to produce poisoning when eaten, may be the result of careless handling of pest control poisons, inclusion of mineral oils used for lubricating machinery, or contamination with lead, arsenic, zinc, or even copper, tin, or aluminium. The pest control poisons will be dealt with in the appropriate paragraph. Mineral lubricating oils cannot be digested by the human system and are suspected of being carcinogenic, they also absorb vitamins from other foods, preventing them from being assimilated by the body. Should there by any likelihood of contamination of foods by mineral oils on machinery, then special vegetable greases should be used. Lead is a cumulative poison which, even when taken in very small quantities, gradually builds up to a toxic quantity, and is mainly associated with lead paints. No machine-parts or utensils, in contact with foodstuffs, should be painted, and it is wiser to use paints other than lead paints in food factories. Arsenic is present in small quantities in many foodstuffs, and so legal limits of 1·4 parts per million for solids (approximately a tenth of this amount for liquids) must not be exceeded. All raw materials must conform to these standards under warranty. Zinc poisoning can result from the use of galvanised vessels as food containers, particularly when used for acid fruits. Copper, tin, and aluminium are toxic, if consumed in large enough quantities, but under normal circumstances they are considered safe. Aluminium is dissolved by strong acids and strong alkalis, so should not be used if there is this likelihood.

Insect infestation

Contributing to both insanitary and unclean conditions are the insect and animal pests, frequently the result of a vicious circle. In hygienic premises, it is very unlikely that these pests will occur and if they do, it is unlikely that they will survive long enough to multiply and cause real concern. Conversely, they will thrive in unhygienic conditions, rendering the foodstuffs more and more liable to contamination of an insanitary or filthy nature.

The common insects, associated with infestation of flour, have already been described, but it should be recalled that these pests can also infest other products and can, and will, breed wherever conditions are suitable. The conditions required are usually warmth, food and moisture, preferably in a dark or undisturbed situation. To prevent infestation, it is necessary to ensure that these conditions do not arise. This is best achieved by regular and careful cleaning, strict and rapid stock rotation, and by destroying any old stock out of use that is likely to be or become infested. If infestation still occurs, many types can be removed by sieving. Those which render the raw materials unsafe through bacterial contamination, should be destroyed by burning to prevent their spreading to other food supplies. Powdered raw materials can be passed through an entoleter on arrival, and by centrifugal force, the insects or eggs are destroyed (but not removed). There are many chemicals capable of destroying insects, and they do so by one or more of three methods. They may be poisonous when eaten, they may dissolve the waxy protective covering of the insect, or the poison may be absorbed through the respiratory system.

Some of the poisons are extremely harmful to humans and domestic animals, and should never be used in food factories.

FACTORY LAYOUT

These are sodium fluoride and thallium sulphate. Boric acid and borax, however, are safe. The type that dissolves the waxy covering of insects is probably the most widely used insecticide, and includes pyrethrins, gamma benzene hexachloride (Gamma B.H.C.), and dichlor-diphenyl-trichlorethane (D.D.T.). They are all available as dusts, sprays, and smokes. Pyrethrins are extracts of certain pyrethrum flowers, which are harmless to man and animals, but fatal to many insects. They are safe to use in food factories. They will kill on application, and at a later stage when contacted by insects, but decompose quickly in the presence of light. Gamma B.H.C., D.D.T. and Dieldrin are fatal on application and at later stages of contact, but are suspected of having cumulative toxic effects to humans and animals, and their use is being actively discouraged. Dieldrin is a synthetic insecticide that is rather more toxic than B.H.C. and D.D.T., but is more effective and more lasting around ovens. It is particularly useful when incorporated in a resin which is painted or sprayed on surfaces around crevices where cockroaches and ants are known to be breeding. Malathion is less toxic than gamma B.H.C., and is very effective in warehouses against weevils and grain beetles. Newer insecticides include Diasanon and Fenitrothion which are not suspected of being stored in the fatty tissues of mammals. The respiratory system poisons are applied in extreme cases of infestation as fumigants, and only by skilled personnel with the permission of the local health office. Methyl bromide is the normal gas used, but ethylene oxide and hydrogen cyanide have been, and still are, occasionally used.

Insecticides may be applied in various ways, but the simplest method is by the use of sprays. Spraying devices come in all sizes and price ranges, from hand sprayers to power-operated sprayers. Most hand-operated sprayers will handle water-based and light oil sprays, but heavy oil sprays require a power-operated machine. Water-based sprays have the widest applications, but oil sprays are necessary for protective spraying on surfaces, and for space spraying against flies. The light oils suitable for manual sprayers are refined paraffin oils which should be odourless, colourless, and tasteless. They are highly inflammable, and great care must be taken when spraying. The heavy oils are much less dangerous and are effective longer, but require power-operated sprayers for application. Dusting insecticides is more satisfactory in some instances, such as along floor crevices. For this purpose, the dust may be sprinkled or blown by pressurised dusting guns. Insecticidal smokes can be used for applying an even coating of insecticide over surfaces both horizontal and, to some extent, vertical, but should not be confused with fumigation, which has greater powers of penetration. Smokes will only protect the outer surfaces of sacks, while fumigants will penetrate through the sack.

The common food factory pests are mites, moths, beetles, ants, cockroaches, and flies. Owing to the bacterial content of their excretion and their smallness, it is particularly important to prevent any development of mites. They prefer moist conditions for breeding, so store rooms should be dry, well ventilated, and regularly cleaned. In the event of infestation by mites, the stock should be destroyed and the store treated with a suitable insecticide. Most infestation can often be sieved from the raw material, but this will not prevent further outbreaks, and as the larval stage is often impervious to insecticides, action must be taken against the adults and the breeding grounds. Cleaning will destroy breeding places, and light oil sprays of pyrethrins are effective against the adult moths. In bad cases, insecticidal smokes may be necessary. Treatment should be repeated at frequent intervals. Beetles and weevils in egg form, are often present in raw materials, so it is necessary to observe strict stock rotation and a quick turnover of susceptible stocks. Entoleter treatment can be useful. Infested materials may be sieved through very fine silks, but extreme cases should be burned or destroyed and the store sprayed with pyrethrins in combination with Malathion which has been found to be very effective against beetles and weevils. Ants are difficult to destroy, owing to their liking for deep, narrow crevices in warm places, but when the lairs are located, the application of a persistent resin around the place where the ants will have to walk, proves very effective. Complete annihilation can be a long, tedious process.

Cockroaches are particularly unpleasant insects, for not only do they look unsightly, they taint food with their excreta and are disease carriers. They frequent warm, dark hide-aways during the day, venturing out for food at night. Cockroaches in bakeries are of two varieties. The small, fully winged German cockroach is often referred to as the steam fly. The large, dark-brown Oriental cockroach, has only short, or almost no, wings. Regular cleaning and filling of crevices, spraying and dusting with a mixture of pyrethrins and Diasanon will control cockroaches, but the most effective method is to paint a persistent resin around the breeding places. This treatment is also suitable for the control of crickets and silverfish. While cockroaches have not evolved in appearance since pre-historic times, they are extremely effective at building up immunity or resistance to insecticides. New insecticides are constantly being developed by specialist firms, whose advice should be sought in all serious cases of infestation.

Flies are potentially the most dangerous pest encountered in food factories, and the most difficult to keep out. Their habits are quite disgusting, and they carry bacteria and disease with them. Every possible effort should be made to eradicate them and to prevent their breeding, not only on the premises, but in the surrounding property as well. They feed and breed in decaying refuse, manure, rotting foods, and decaying faeces, most of which are certain to be infected with pathogenic bacteria and possibly with disease-producing organisms and the eggs of parasitic worms. The fly visits these places to lay eggs and to feed, and then returns to foodstuffs in the factory, carrying bacteria not only on its feet, but also in its digestive system. While on the foodstuffs in the bakery, it vomits saliva on to the food, rendering the food assimilable, and it defaecates at the same time as it eats. Thus the food becomes infected three times over. If the foodstuffs and conditions are particularly suitable, bacteria and other organisms can multiply and develop at an alarming rate. Flies can be the bearers of diseases such as cholera, dysentery, and typhoid and paratyphoid fevers. Trichinosis is an illness caused by a parasitic worm, the egg of which flies can carry and transmit to humans.

To prevent infestation by flies, all potential breeding grounds must be destroyed and disinfected. To prevent flies from entering the factory, fine mesh screens should be placed over all opened windows. Air-curtains at doors which are frequently opened, stop flies entering (an air-curtain is a strong current of air blown across the door opening), but do not impede the passage of operatives or materials. Air conditioning precludes the necessity to open windows, and all the air entering the building is filtered. All waste food should be cleared from the factory as often as possible, and pig-food and refuse areas should be kept disinfected and clean. The waste should be cleared as often as possible, and all bins must have well-fitting lids. All foods in the factory should be kept covered wherever possible, and the most suitable media for bacterial growth should be kept at refrigerated temperatures. Malathion is an effective fly-killer, and can be used as a powder round refuse areas, and as a spray on ceilings and walls, where it will retain its efficiency for several weeks. A 'knock down' spray of pyrethrins can be effective against flying adults, but does not prevent future infestation. The use of insecticides (and rodenticides) must always be a cause for concern in food premises, and the utmost care must always be exercised. Electrical devices are available which attract flying insects by means of a light. As the insect alights on the grid surrounding or adjacent to the source of attraction, it is electrocuted and falls into a tray provided.

Animal infestation

The common animal pests to be found in food factories are rats, mice, and birds. Dogs and cats sometimes gain access, but should be excluded. All these animals are capable of being disease carriers, and some will foul foodstuffs and cause damage.

Rats are the worst of these pests, and cause incalculable damage to buildings, raw materials, and containers, apart from infecting the raw materials and finished products with the bacteria and disease organisms they carry. Diseases for which rats can be responsible are bubonic plague, typhus, dysentery, some forms of jaundice, poliomyelitis, and various forms of food poisoning. Not only do they pick up germs during their travels through filth and refuse, but they suffer from diseases which they can pass on to humans. Rats are largely nocturnal and furtive creatures, and may be present in some strength, without actually being seen. They can be readily detected by the damage they do to sacks and bags containing raw materials, and to floors, doors, and walls, by gnawing and burrowing. Much of the damage done is not only to gain access to food or premises, but also to keep their teeth sharp and short. Other signs, indicating the presence of rats, are pellets of excreta, greasy smears along walls and girders where they rub their fur when passing, and tracks caused by feet and tails through flour dust.

As with the control of insect pests, prevention is better than cure, and to prevent infestation by rats, the premises must be carefully maintained to keep them out. Doors must fit closely; drains, manhole covers, rainwater pipes, and air vents, must be kept so that rats cannot use them as entries. Spaces round pipes that pass through walls should be blocked. Deep foundations will prevent rats burrowing under from outside. All heaps of junk, refuse, and unused machinery should be cleared away regularly, as these make ideal breeding grounds. Foodstuffs must be cleared, and pig-food and waste should be in bins with tight fitting lids. Preferably, the bins should be raised from the floor, so that the area can be swilled daily. Rats are suspicious of new objects which appear in their usual runs, so it is necessary to accustom them to traps and poisoned baits before they will feed. Break-back traps should be placed with the treadle across the run and left baited with oatmeal, but not set for a few days. When the bait starts to disappear, the trap should be set. Gloves should be worn, as rats are suspicious of human smells—perhaps with good reason. Trapping is not the most efficient way of dealing with rats, but at least the score of successes is evident. This is not so with poisoning, but when successful, there will be no more traces to be found. Poisoning can be accomplished by the use of acute poisons which kill after one meal, and chronic poisoning which builds up gradually. With acute poisons for rats, unpoisoned baits should be set for a few days followed by the same bait including the poison. Suitable poisons are zinc phosphide, arsenious oxide, or Antu. It is safer for these poisons to be placed in short lengths of drainpipe, to prevent domesticated animals gaining access to them. Each poisoned bait should be noted, so that it can be collected afterwards if not taken by the rats. Warfarin is used as the chronic poison, and pre-baiting is not necessary. Rats come and go over fairly wide areas, and it is desirable, therefore, to campaign against them in concert with neighbouring property owners.

Mice are also extremely undesirable creatures on premises where foodstuffs are manufactured. Similar treatment should be given to mice as that already outlined for rats, except that mice are inquisitive by nature and will investigate traps and baits on their first appearance.

When using poisons against rodents, due regard must be given to the position and chance of contamination. From this aspect in food premises, Warfarin is probably the safest and wisest to use. In some cases, however, rodents have developed immunity to Warfarin, and it should be replaced with alpha chlorolose and specialist advice should be sought. In all cases of rodent infestation, the local authority must be advised.

Bird pests in food factories consist largely of sparrows, but include starlings and pigeons. Pigeons are known carriers of bacteria of the salmonella group and, in addition, can foul foodstuffs by their indiscriminate defaecating, and by shedding feathers and nesting materials, and by disturbing dust. Their nests are frequently infested by moths and beetles. If effective fly screening and rat control has been carried out, birds should not gain access, except through open doors where strip curtains could be used. It is illegal to poison birds, so those in the building should be caught in cages or traps and disposed of humanely (unless they are protected birds), or taken away and released.

GOOD HOUSEKEEPING

Good housekeeping, is the term given to hygiene concerned with premises and plant. Its observance is greatly facilitated by good layout, and design and planning of both premises and plant. The premises should be of materials which are hard-wearing and easily cleaned. There should be no corners and crevices, no dark and dingy places, no heaps of rubbish, refuse, or junk. Machines, cupboards and pipes should be away from walls and raised from the floor, so that they can be cleaned easily behind and beneath. Ledges and pipes which collect flour and dust must have easily cleaned surfaces and be accessible. Raw materials should be readily moved for cleaning behind and under, stocks should be rotated strictly and quickly, store room conditions should be ideal. Suction cleaners should be used to remove all dust and dirt, rather than sweeping, which tends to push the dirt into crevices. Efficient ventilation is necessary to reduce temperatures and humidity, and so discourage development of bacteria, moulds, and insect pests. Waste bins in the factory should be emptied frequently, and the area for waste should be kept well swilled and disinfected. The bins should be raised from the floor and have well-fitting lids. When empty, they should be washed and disinfected. Various types of bins should be reserved for various types of waste, and all burnable waste should be burned as it accumulates. Other forms of waste must be disposed of quickly and regularly. The design of plant and equipment is very important, and should be such as not only to improve the appearance of the machine, but to render the inaccessible parts more readily accessible while still remaining safe. Panelling down to the floor makes plant attractive and easily cleaned on the outside, but makes dark, undisturbed conditions, ideal for insect and animal breeding and feeding. It is not sufficient to have a clear space under the equipment. The panels should be easily removable, when the machine is not running, so that the working parts are convenient for regular cleaning. A great deal of foodstuffs can accumulate under the machine, behind the guards or panels.

In order to promote good housekeeping, the responsibility should belong to an executive engaged chiefly for this purpose. It is his duty to see that each section carries out the regular and routine systems of cleaning of their own department. Areas of common ground and 'no-man's-land' should be cleaned by a hygiene gang under the direct control of the hygiene executive. It is his duty to supply and control cleaning equipment and detergents. The choice of detergents can vary, according to the requirements and limitations of the particular job in hand, but detergents should not be toxic, not impart unpleasant flavours, not be affected by hard-water, nor should they be corrosive. They should have good wetting and emulsifying powers, the ability to hold solid particles in suspension, be readily soluble in water and, wherever possible, they should be bactericidal.

In order to clean small equipment, special sinks should be provided in which the utensil can be washed in water at 49-54°C (120-130°F), containing a suitable detergent. When the utensil is clean, it should be allowed to drain and then immersed in a second sink containing water at a temperature exceeding 82°C (180°F), which may contain a trace of disinfectant, such as sodium hypochlorite. The utensil is hot enough when removed from the sterilising sink to dry quickly without being touched. Wiping with a cloth only puts back bacteria. Hands must not be washed in utensil sinks, and vice versa. Separate wash hand basins must be provided.

Floors should be cleaned with suitable detergents in hot water. It may be necessary first, to scrape the floor in areas where raw materials and dough are liable to be spilled. Suction cleaners are invaluable in dusty areas. Where floors are regularly swilled, there should be no gutters covered with gratings, as these are easily neglected and soon become ideal breeding places. If grids are necessary for the removal of water, they should be kept scrupulously clean and regularly disinfected.

PERSONAL HYGIENE

In any factory where food is prepared, personal hygiene, from the youngest apprentice to the managing director, is of the utmost importance. Cleanliness of the highest order must be demanded, but it is a principle that is not easily enforced. To achieve the highest standards, it is necessary to appeal to the operatives' consciences, and to make the operatives conscious of the dangers and implications of unhygienic habits and practices by displaying notices and posters, and by talks and film shows. However strongly hygiene is emphasised, propaganda is quite pointless without adequate facilities being provided. Wash hand basins, complete with soap, nail brush, hot water, and drying facilities, must be placed where they are readily accessible after using the toilet, and where work is likely to dirty the hands. The provision of towels which are used by everyone and quickly become dirty, is quite worthless. Disposable towels or hot air drying appliances should be provided. All operatives should wear clean protective clothing, which should be changed when soiled. They should refrain from putting their hands and fingers to mouth or nose, and a clean handkerchief should be carried for nose-blowing, after which the hands should be washed. Finger-nails should be kept short and well scrubbed. Finger-nail biting should be discouraged. Carriers of diseases must not be employed, and any suffering from infections of the respiratory system, boils or septic wounds or intestinal disorders, should not be permitted to handle food. Spitting must not be allowed, and smoking forbidden, in any place where food is handled or prepared. All staff should pay routine visits to a dentist.

COST AND EFFECTS OF HYGIENE

The cost of an efficient and effective hygiene system will be considerably higher than no system at all, but can be very inexpensive when compared with the benefits that accrue. The obvious advantages are that products manufactured on hygienic premises, by a hygiene-conscious company, will be free from pathogenic organisms, toxic substances, and fragments of insects or filth. The presence of any of these will result in law suits and prosecutions and consequently, adverse publicity, high expense in costs, fines and compensation, and loss of prestige and custom. Loss of business may even result in eventual bankruptcy of the company, and unemployment for the operatives. Less obvious advantages of an effective hygiene system are in the reduction of accidents, and in a high standard of morale among the employees.

All factors contributing to the good of the morale of the employees should be pursued and encouraged. Well planned, pleasant hygienic working conditions, are excellent morale boosters. During meal breaks, the employees should be able to relax, and be completely detached from their daily work routine. Canteen and rest-room facilities should be bright and cheerful, with provision for relaxing activities, preferably of a not-too-boisterous nature. The meal break activities can well be extended to after-work or weekend social groups or clubs, fostering the *esprit de corps* or 'togetherness' of the company. Good morale can lead to less absenteeism, better time-keeping, and a small turnover of operatives, all of which lead to a higher standard of work and increased productivity. The manufacture of biscuits, or any other product, depends upon the complete cohesion of a team. If the team lacks confidence in each other, and particularly in the captain or team manager, few games will be won. A successful and happy team, with faith in each other, will not only win the game, but the title as well.

Bibliography

Bennion, E. B. (1967) Breadmaking, Its principles and practice, 4th edition, London, Oxford University Press.

Bennion, E. B., Stewart, J., and Bamford, G. S. T. (1966) Cakemaking, 4th edition, London, Leonard Hill Books.

Fance, W. J. (1966) The Students Technology of Breadmaking and Flour Confectionery, 2nd edition, London, Routledge and Kegan Paul.

Kent Jones, D. W. and Amos, A. J. (1967) Modern Cereal Chemistry, 6th edition, London, Food Trade Press Ltd.

Matz, S. A. (1968) Cookie and Cracker Technology, Westport, Connecticut, AVI Publishing Co., Inc.

Munn Rankin, W. and Hildreth, E. M. (1966) Foods and Nutrition, 9th edition, London, Allman & Sons Ltd.

Urie, A. and Hulse, J. H. (1952) The Science, Raw Materials and Hygiene of Baking, London, Macdonald & Evans Ltd.

Index

Accelerated freeze dried egg (A.F.D.), 57
Accidents and fatigue, 274
Acid calcium phosphate (ACP), 15, 45
Acid value of fats, 176
Acidity
  of dough, 110, 112
  of flour, 14, 15, 175
  of milk powder, 176
Activated dough development, 113
Additives
  in flour, 16
  legal aspects of, 16
Aerating agents, 42
Aeration
  biological, 42
  chemical, 43
  mechanical, 42
  lamination, 42
Agar-agar, 70
  jellies, 156
Air
  blowers, 229
  curtains, 281
Albumen, 6, 9, 57, 110
Alcohol, 110
Alginate, 71
All-in method, 105
Allspice, 84
Almond(s), 66
  flavour, 88
  paste, 66
Aluminium foil, 262
Alveograph, 165
Amaranth, 98
Ammonium
  bicarbonate, 43
  carbonate, 43
  sulphate, 15
Amylase (α and β), 110
Angelica, 64
Aniline dyes, 97
Animal
  fats, 22, 23
  infestation, 282
Annatto, 96
Arachis, 27

Arrowroot, 21
  biscuits, 145
Artificial colours, 97
Ascorbic acid, 113, 114
  in flour, 174
Ash in flour, 9, 10, 12
  determination, 171
Aspiration, 7, 8
Autolysed yeast, 82
Automation, 256

Bacteria, 275
  effect of pH on, 49
  in eggs, 58
  and fermentation, 112
  in flour, 110
Bacterial contamination, 275
Bacterial enzymes, 16
Baker's chocolate, 74
Baking
  biscuits, 146, 233
  control, 182
  and ovens, 230
Barbados sugar, 34
Barcelonas, 67
Barley, 19
Base for marshmallow, 132
Beet sugar, 34
Benzaldehyde, 89
Birds in factories, 284
Biscuit
  baking, 146, 233
  classification, 103
  cooling, 234
  effect of pH on, 48
  flow, 40, 48, 183
  formation, 222
  formulae, 127
    guide to use, 127
  grinding, 250
  packaging, 258
  spread, 40, 48, 183
  structure, 103, 233
  texture, 40, 104
    meter, 180
  waste, use of, 189

INDEX

Blackjack, 96
Bleaches, 14
Bleaching of flour, 13
Blending of flour, 202
Bloom strength, 70
Blue, brilliant, 98
Bourbon, 132
Bran, 5, 7, 10
  in flour, 174
Branny particles, 11, 14
Brazil nuts, 67
Break rollers, 7
Bromates in flour, 15, 174
Budding of yeast, 106
Bulk storage, 201
Butter, 23
  composition, 23
  flavour, 90
Buttermilk, 52

Calcium
  carbonate, 17
  sulphate, 15
Candied peel, 63
Candling of eggs, 56
Cane sugar, 34
Caramel, 96
Caraway seeds, 85
Carbohydrates, 38
Carbon dioxide, 110
Cardamom, 85
Cardboard, 260
Carmine, 95
Carotene, 13, 15
Carragheen, 71
Cashew nuts, 68
Cassia, 83
Celery seeds, 85
Cellophane, 260
Cellulose, 5, 10, 38
  film, 260
Cereal
  enzymes, 16
  products, 1, 18
Checking, 185, 234
  prevention of, 186, 234

Cheese, 54
  crackers, 142
  filling cream, 159
  storage, 54
Chemical contamination, 277
Cherries
  crystallised, 63
  glacé, 63
Chipboard, 260
Chlorides in flour, 174
Chlorine, 15, 173
Chlorine dioxide, 15, 173
Chloroform test, 173
Chlorophyll, 96
Chocolate, 73
  bloom, 78
  flavour, 91
  -flavoured shell, 132
  handling, 76, 206, 246
  production, 74
  tempering, 75, 247
  tests, 179
  uses of, 79
Chorleywood process, 113
Cinnamon, 83
Citric acid, 44
Citron peel, 63
Citrus peels, 63
Cloves, 83
Cochineal, 95
Cockroaches, 18, 280
Cocoa, 73
  bean, 73
  butter, 27, 73
  products, 73
  uses of, 79
Coconut, 65
  oil, 25
  ring, 138
Coffee extract, 91
Colour
  artificial, 97
  of flour, 10, 172
  natural, 95
  regulations, 97, 99
  use of, 94
Colouring materials, 94
Conditioning of wheat, 7

INDEX

Conduction of heat, 232
Control
  process, 181
  quality, 161
Convection of heat, 232
Conveyors, 252
Cookie formulae, 137
Coriander, 85
Corn syrup, 37
Cottonseed, 28
Couverture, 74
Craeta preparata, 17, 172
Cream
  biscuits, 131
  cracker(s), 140
    dough, 116
    dust, 141
    lamination, 116
  fillings, 156
  handling, 207
  icing, 160
  of tartar, 45
  powder, 46
Creaming
  machines, 243
  method, 104
Crystallised flowers, 64
Crystallised fruits, 64
Currants, 59
  types of, 60
Curry powder, 85
Cutting machines, 222
Cylinder test, 12
Cysteine hydrochloride, 114
Cytase, 5, 10

Dairy products, 50
Dates, 62
Demerara sugar, 34
Detergents, 285
Development, 193
Dextrins, 5, 20, 38, 110
Dextrose, 36, 110
Diastase, 5, 20, 110
Diastatic activity, 15, 20

Dielectric heating, 238
Digestive (sweetmeal), 133
Disaccharides, 38
Dough
  brake, 220
  depositor, 227
  development, 113
  handling, 217
  mixing, 104, 114
  sheeter, 219
  temperatures, 114, 118, 121
  test, 13
Dried fruit, 59
  in biscuit doughs, 61
Dust explosions, 204
Dyox, 15, 173

Egg(s), 55
  dried, 57
    A.F.D., 57
  frozen, 56
  pasteurisation of, 58
  preservation of, 56
  products, 55
  substitutes, 58
  test, 56
  whites, 57
  yolks, 57
Electronic ovens, 238
Emulsifiers, 30, 32
  in filling creams, 156
Endocarp, 4
Endosperm, 5, 7
Enrobers, 247
Enzymes, 109
  fungal, 15
Enzymic action, 5, 10, 110
  in fats, 31
Epicarp, 4
Epidermis, 4
Ergot, 17
Erythrosine, 98

INDEX

Essences, 86
  natural, 86
  requirements of, 91
  synthetic, 87
  uses of, 92
Essential oils, 22, 86
Esters, 87
Ethyl alcohol, 110
Eugenol, 88
Explosions dust, 204
Extensibility, 16
Extensimeter, 165
Extensograph, 165
Extensometer, 169
Extraction rate, 7
Extracts for flavour, 86

Factory
  fabric, 273
  layout, 269
Farinograph, 162
Fat
  bloom on biscuits, 187
  bloom on chocolate, 75, 77, 78
  chemical nature of, 29
  in flour, 9
  handling, 205
  Kreis test, 176
  processing, 205
  in puff doughs, 118
  slip melting point of, 177
  specific gravity of, 178
  storage of, 204
Fatigue and accidents, 274
Fats
  and oils, 22
  acid value of, 176
Fatty acids, 29
Faults in biscuits, 183
Fermentation, 105
  acid, 110
  effect of pH on, 48
  panary, 109
  results of, 112
  tolerance, 12

Fermented doughs, 105
  formulae of, 139
Fibre in flour, 10
Figs, 62
Fig bar formation, 227
Filbert nuts, 67
Filling creams, 156
  savoury, 158
Flavour, 81, 92
  by fermentation, 112
  effect of pH on, 48
  in cream biscuits, 131
  in marshmallow, 152
Flavouring materials, 81
Flies, 281
Flour, 1
  acidity, 14, 175
  analyses, 9
  bleaching, 13
  blending, 202
  characteristics, 2, 8
  colour, 10, 172
    grade, 172
  composition, 8
  constituents, 9
  contamination, 17
  fractions, 8
  heat treatment of, 14
  Manitoba, 12
  mite, 18, 280
  moisture, 9, 12, 171
  moth, 17, 280
  quality, 10
  sprinklers, 229
  storage, 18, 201
  strength, 6, 11
  tests, 10-16, 161-176
  treatment, 13-16
    detection, 172
  water absorbing properties of, 11
  weevil, 17, 280
  wheatmeal, 7
  white, 7, 10
  wholemeal, 6
Flying sponge and dough process, 115, 141
Foil, 262

INDEX

Formulae
  of biscuits, 127–147
    guide to use, 127
  of products other than biscuits, 148–160
Fourths, sugar, 35
Fractionation, 8
Fructose, 36, 38, 110
Fruit and nuts, 59
Fruit
  biscuit, 134
    (hard dough), 146
  crunch, 137
  flavours, 90
  in biscuits, 61
  puff, 143
Fuels, 237
Fungal enzymes, 16
  in flour, 175
  in hard doughs, 122
  in puff doughs, 120

Garibaldi biscuits, 146
Gas packaging, 266
Gauge rolls, 219
Gelatine, 69
  in marshmallow, 70
Germ (wheat), 4, 5, 7, 10
Ginger, 82
  biscuits, 136
  nuts (ginger snap), 135
  root, 64
Glassine, 260
Globulin, 6, 110
Glucono-delta-lactone (G.D.L.), 46
Glucose, 36, 37, 38, 110
  commercial, 37
Gluten, 9
  effects of fermentation on, 112
  estimation, 170
  -forming proteins, 6, 9
  quality, 11
  quantity, 11, 12
  stability, 14

Glutenin, 6
Glycerides, 29, 30
Glycerine (Glycerol), 29, 30
Golden syrup, 36
Good housekeeping, 284
Grease-proof paper, 260
Green S., 99
Grist, 8
Groundnut oil, 27
Groundnuts, 67
Gum(s), 72
  arabic (acacia), 72
  glaze, 72

Handling of raw materials, 201
Hard dough
  biscuits, 144
  lamination, 122
  production, 105, 121
Hazelnuts, 67
Heat
  sealing, 262, 265
  transference, 231, 233
  treatment of flour, 14
High frequency baking, 238
Hollow bottoms (of biscuits), 184
Honey, 37
Humidity
  chocolate room, 78
  dough room, 115
  oven, 237
  paper stores, 263
Hydrogenation, 25, 30, 31
Hydrogen ion concentration, 47
Hydrolysis of carbohydrates, 38, 39
Hygiene, 269, 275
  costs, 286
  personal, 286

Icing(s), 159
  cream, 160
  equipment, 249

INDEX

Impact milling, 8
Indigo carmine, 98
Infestation, 17, 278, 282
   of flour, 17
Insectides, 278
Insect(s), 17
   infestation, 278
Insulators, 232
Inversion, 39
Invertase, 39, 110
Invert sugar, 36, 39, 110
Iodates in flour, 174
Irish moss, 71
Iron in flour, 17
Isinglas, 71

Jam(s)
   and jellies, 153
   depositing, 244
   handling, 207
Jellying agents, 69

Kernel paste, 66
Kjeldahl, 170
Kraft paper, 261
Kreis test, 176

Lactic acid, 52, 54
Lactose, 38, 39
Laevulose, 36, 110
Laminated wrapping materials, 262
Lamination, 42
Laminators, 220
Lard, 24
Larvae (moth), 17, 18
Lauric acid, 30
Leaf gelatine, 69
Lecithin, 21, 32
   in egg, 57
   in filling creams, 156

Lemon
   flavour, 89
   peel, 63
Lincoln (cream), 130
Linoleic acid, 31
Linolenic acid, 31
Lintner, 20
Lipase, 110
Liqueurs, 91

Mace, 83
Machining control, 181
Machine room equipment, 219
Magnesium sulphate, 15
Maize, 19, 28
Malt
   extract, 20, 38
   (dried), 20
   flour, 20
   production, 19
   products, 20
   sugar, 10, 20
   syrup, 38
Maltase, 110
Maltose, 10, 20, 38, 39, 110
Maple syrup, 38
Margarine, 28
   in puff doughs, 118
Marie biscuits, 144
Marshmallow, 150
   aeration, 152
   biscuit base, 132
   biscuit conditioning, 153
   equipment, 244
Marzipan, 66
Maw seeds, 85
Meal
   germ, 7
   wheat, 7
   whole, 6
Mechanical dough development, 113
Melloene treatment, 16, 122
Metal detectors, 252
Mice, 282
Mildew, 17

INDEX

Milk, 50
  composition of, 53
  condensed, 50
  dried, 51
  powdered, 51
    acidity of, 176
  reconstitution of, 51
Milling
  flour, 6
  impact, 8
  roller, 6
  stone, 6
  wheat, 6
Mineral
  improvers, 12, 15
  oils, 22
  salts in flour, 9, 10, 173
Mites, 18
Mixed spice, 86
Mixers
  continuous, 210, 214
  horizontal, 209, 212
  miscellaneous, 210, 216
  reciprocating arm, 210, 214
  vertical, 209, 210
Mixing
  control, 181
  methods, 104, 114
  room equipment, 209
Moisture
  determination, 171
  in biscuits, 146
  in filling creams, 158
  in flour, 9, 12, 171
  in jam, 153, 179
Molasses, 36
Monkey nuts, 27, 67
Monocalcium phosphate, 45
Monosaccharides, 38
Monosodium phosphate, 46
Moths in flour, 17
Moulders, rotary, 224
Mould
  effect of pH on, 49
  in cheese, 54
  spores, 18
MSAT film, 260

Mucor mucedo, 18
Muscovado sugar, 34
Mustard, 85
MXXT film, 261

Natural colours, 95
Nice biscuits, 130
Nicotinic acid, 17
Nitrogen peroxide, 14, 172
Novadelox, 14
Nucellar layer, 5
Nutmeg, 83
Nuts
  and fruits, 59
  and nut products, 65

Oats, 18
Offals (flour), 7
Oil
  extraction, 25
  hardening, 25
  in flour, 9
  refining, 25
  spray machines, 251
  wheat, 13
Oils
  and fats, 22
    chemical nature of, 29
  essential, 86
Oleic acid, 30
Olive oil, 28
Orange
  flavour, 89
  peel, 63
Osborne biscuits, 144
Oven bands, 230
  tension on, 231
Oven(s)
  and baking, 230
  electronic, 238
  heating systems, 235
  turbulence, 237

INDEX

Oxidation
  of flour, 13, 15
  of fats, 30, 31, 32, 206
Oxidising agents in dough, 113

Packaging, 258
  machines, 263
  materials, 259
  standards, 182
Palmitic acid, 30
Palmitoleic acid, 30
Palm kernel oil, 26, 27
Palm oil, 26
Panary fermentation, 109
Paper, wrapping, 259
Paprika, 85
Pasteurisation
  of coconut, 66
  of eggs, 58
Peanuts, 27, 67
Pectin, 71
  in jam, 153
  in jellies, 155
Peels, 63
Pegmill, 8, 250
Pekar test, 10
Pepper, 84
  cayenne, 85
Peppermint flavour, 90
Personal hygiene, 286
Persulphates in flour, 14, 174
Petit Beurre biscuits, 145
pH
  determination, 48
  effect on biscuits, 48
  scale, 47
Phosphates in flour, 15, 173
Photosynthesis, 38
Pieces, sugar, 35
Pillow packs, 266
Pimento, 84
Piping jelly, 70
Pistachio nuts, 67
Plastics trays, 260, 266
Plumule, 5

Pneumatic handling, 202
Polypropylene film, 261
Polysaccharides, 38
Polythene film, 261
Poppy seeds, 85
Potassium
  bromate, 15, 114
  phosphate, 15
Praline paste, 67
Preparation of ingredients, 207
Process control, 181
Processing standards, 182
Production methods, 103
Protease, 110
Protein
  displacement, 8
  estimation, 170
  gluten-forming, 9, 11
  insoluble, 6
  soluble, 5, 9, 11
Proteolytic enzymes, 16, 110
Puff doughs, 117
  aeration of, 42, 120
  faults in, 120
  formulae of, 143
  and fungal enzymes, 120
  lamination of, 119, 222
  Scotch method, 222
Pumpable shortening, 205
PVDC film, 261

Quality
  control, 161
    of flour, 161
    records, 187
  of flour, 10

Radiation, 232
Radicle, 5
Raisins, 61
Rancidity, 30, 31, 176
Rats, 282

INDEX

Raw materials
  handling, 201
  preparation, 207
  storage, 201
Records, 187
Reducing agents in doughs, 114
Reduction rollers, 7
Refractometer, 179
Rennet, 54
Research testing unit, 166
Rice, 19
Rice weevil, 17
Rich tea biscuits, 144
Rodent infestation, 282
Roller milling, 6
Rollers
  break, 7
  reduction, 7
Rotary moulders, 225
  problems, 225
  web treatment, 226
Rout press
  biscuit formulae, 139
  machines, 226
Rust on wheat, 17
Rye, 19

Saccharomyces cerevisiae, 106
Saffron, 96
Salmonella, 276
  in coconut, 65
  in egg, 58
Salt, 81
  sprinkler, 227
Savoury
  crackers, 142
  filling creams, 158, 159
Scrap use, 189
Scutellum, 5
Semi-sweet hard doughs, 121
  formulae of, 144
Setting agents, 69
Shells for creaming, 131, 145
Shortbread, 129
Shortcake biscuits, 128

Shortening,
  pumpable, 205
  purpose in doughs, 29
Slip melting point, 177
Smut on wheat, 17
Sodium
  acid pyro-phosphate, 45
  alginate, 71
  bicarbonate, 43
    action on gluten, 6, 44, 183
  chloride, 81
  m-bisulphite, 16, 122
  phosphate, 15
  sulphate, 15
Soft dough
  biscuits, formulae of, 128
  production, 103
Soya
  beans, 28
  flour, 21, 55
Specific gravity of fats, 178
Spices, 82
  extracts, 86, 90
  mixed, 86
Sponge and dough process, 115, 140
Spores, 17, 18
Sporulation (of yeast), 107
Stabilisation of flour, 14
Stabilisers and emulsions, 30
Standards, 181–183
Staphylococci, 276
Starch, 5, 8, 11, 38
  cells, 8, 110
Stearic acid, 29
Stearin, 29
Stone milling, 6
Storage
  control, 182
  of flour, 18, 201
  of raw materials, 201
Straight dough process, 114, 140
Structure of biscuits, 103, 233
Sucrose, 10, 34, 38, 110
Sugar, 34
  beet, 34
  bloom, 77, 78
  cane, 34
  chemical nature of, 38

INDEX

effects in baked goods, 39, 103, 183
grades, 35
grinding, 250
handling, 203
in flour, 9, 10, 11
products, 36
refining, 34
refractometer, 179
solids in syrups, 179
sprinkler, 228
storage, 203
tests, 178
types, 40
Sulphates in flour, 15, 174
Sulphur dioxide, 16
Sultanas, 60
Sunset yellow, 98
Sweetening agents, 34
Sweetmeal (Digestive) biscuits, 133
Synthetic essences, 87
Syrups, 36, 37
    handling and storage of, 206

Tartaric acid, 44
Tartrazine, 97
Tea biscuits, 144
Tempering of chocolate, 75
Testa, 5
Tests
    on biscuit texture, 180
    on chocolate, 179
        viscosity, 179
    on egg, 56
    on fats and oils, 176–178
    on flour, 10–13, 161–176
    on milk powder, 176
    on sugar, 41, 178
Toughness of dough, 184
Treacle, 36
Trinidad sugar, 34
Triticum vulgare, 1
Trouble shooting, 183
Turbulence in ovens, 237
Turmeric, 96

Vanilla, 88
Vanillin, 87, 88
    ethyl, 88
Vegetable oils, 22, 25
Viscosity tests, 179
Vitamin B1, 17
Vol, 43, 128
Vostizza currants, 60

Wafers, 148
    aeration of, 149
    equipment, 241
    filling cream, 150
Walnuts, 66
Wash-over brush, 228
Waste disposal, 189
Water
    -absorbing properties, 11
    absorption meter, 169
    biscuits, 142
        handling of, 117
    handling, 206
    in flour, 9, 12
Waxed paper, 260
Weevils, 17
Whale oil, 24, 25
Wheat, 1
    characteristics, 2
    colour, 2
    composition, 3
    conditioning, 7
    diseases, 17
    germ, 5, 7, 10
    grist, 8
    hard, 3
    Manitoba, 12, 13
    meal flour, 7
    oil, 13
    place of growth of, 1
    section, 4
    soft, 2
    spring, 2
    sprouted, 174
    starch, 5, 8
    time of planting of, 2

Whey, 54
White flour, 7, 10
Wholemeal flour, 6, 16
Wire-cut biscuits, formulae of, 137
Wire-cut machines, 226
Wrapping
    machines, 263
    materials, 259
        sealing of, 262
        storage of, 263

Yeast, 42, 106
    aeration, 42
    autolysed, 82
    dried, 109
    extracts, 82
    food, 10
    reproduction, 106
    storage, 109

Zymase, 110